100 JOBS IN THE
ENVIRONMENT

100 jobs in the environment

Debra Quintana

Macmillan • USA

MACMILLAN
A Simon & Schuster Macmillan Company
1633 Broadway
New York, NY 10019

Book design and production by Sandy Bell
100 Jobs in the Environment is produced by becker&mayer!, Ltd.

MACMILLAN is a registered trademark of Macmillan, Inc.

Library of Congress Cataloging-in-Publication Data

Quintana, Debra.
    100 jobs in the environment / Debra Quintana.
        p.   cm.
    Includes bibliographical references.
    ISBN 0–02–861429–1
    1. Environmental sciences—Vocational guidance.   I. Title.
GE60.Q56   1996
363.7'0023—dc20   96–25511
                                           CIP

10 9 8 7 6 5 4 3 2 1

Printed in the United States of America

ACKNOWLEDGMENTS

*It seemed like such a simple idea, but this book wouldn't have been possible
without the more than one hundred people who gave their time
to talk about their jobs. They were all eager to pass along
the best advice and encouragement for others to follow in their footsteps—
the most credit must be handed to each of them.
Many professional associations were also extremely helpful in
making the connections to just the right person to interview, and offering
overall perspectives to careers regarding the qualifications and prospects
included in the job descriptions.
Thanks also to Betsy Dowling, a terrific researcher and writer who
sought out, conducted, and wrote up several of the interviews.
A special thanks to my husband, Louis Young, a journalist whose guidance and
encouragement I relied upon to complete this project.*

100 JOBS

CONTENTS

INTRODUCTION

TO SAY YOU want a job in the environment sounds very well meaning and altruistic, but what are the career choices? Can you really make a living at it?

This book answers those questions, offering examples of the types of traditional as well as new jobs in the environmental field. It offers solid, concrete advice for those planning environmental careers, those changing careers, and those ready to launch into the job market. Profiles of people actually doing this work take you through a day in the life of these professionals, providing firsthand information about what it takes to get these jobs and what it means to actually do them.

The book also tries to spell out the hottest job opportunities: where to find them and what training and skills are required.

The environmental industry is a young, dynamic, and demanding field offering a myriad of challenging but rewarding careers. It sprang forward in the late sixties with concerns regarding pollution, toxic waste, and other environmental dilemmas. The environmental movement did not formally take hold, however, until the seventies, with the implementation of the Clean Air Act and the creation of the Environmental Protection Agency. Today, colleges and universities offer formal degrees in environmental science and countless other specialties.

In the late 1980s and early 1990s, the field boomed with opportunity as government agencies and private industry expanded to hire specialists in environmental protection, conservation, and remediation. Many large private industries created internal environmental safety offices to monitor their own operations, and to assure compliance with governmental regulations.

Although the environmental field continues to expand as we approach the millennium, it has slowed in some respects because of government funding and corporate budget cutbacks. Opportunities in the field depend largely on the political climate and business and regulatory conditions. At the same time, those cutbacks have sparked growth in some careers such as fund-raising. Recent laws and regulations have also created new jobs such as organic farmers and food inspectors.

What's exciting about the environmental field is that while there is a call for technical jobs—engineers, biologists, chemists, geologists—the area offers opportunity for those who want to carve a new niche in the industry. Entrepreneurs are creating environmentally friendly "green" products, and traditional jobs are incorporating ecological aspects and creating new specialties, like green architecture and green industrial design.

The boom in environmental jobs continues to occur in private consulting firms that do everything from testing and sampling to remediation and restoration. But there is still a great opportunity for work in the public sector and a great need for ecological watchdogs—activists, grassroots organizers, lobbyists and environmental lawyers, and law enforcement officers. There's also a growing job market in the areas of education and communication.

While the various careers require different qualifications, nearly every job calls for the ability to write well and to use computer skills. Many of the technical jobs require writing reports that are an essential part of the work. Communication skills are critical in conveying data results and analysis. Whether it was a scientist, entrepreneur, or park ranger, nearly everyone pointed out the importance of having good writing skills.

As for salaries, they run the gamut. Environmental jobs notoriously offer low salaries. But for most people who enter the environmental market, getting rich is not the primary motivation. Don't be discouraged. There are many well-paying environmental jobs that offer a career track with promotion and advancement. Many of those jobs are in the private sector, especially with consulting companies.

In today's world, you don't run into many people who talk about how much they love their work. That's not the case with environmental jobs. All of the people interviewed in this book (except, perhaps, for the woman who works in sewers) love their career choices.

Whatever the career choice, environmental work demands a conviction and dedication that can't be measured strictly in dollars or work hours. The reward is making the world a better place.

—Debra Quintana

100 JOBS IN THE ENVIRONMENT

1. Wildlife Rescuer/ Stranding Coordinator

description: A stranding coordinator's job is to rescue stranded or injured coastal mammals and marine wildlife, such as seals, whales, dolphins, and sea turtles, so they can be nursed back to health and re-released into their coastal or ocean habitats. Stranding coordinators usually work for nonprofit wildlife animal rescue organizations. When stranding coordinators are not rescuing animals, their job is to feed, care for, and administer medical treatments to rescued animals, as well as to maintain routine paperwork. Expect to work long hours, including weekends, and to be called out to work in the middle of the night to help save an animal.

salary: Stranding coordinators make between $20,000 and $25,000 a year. It's not a 9-to-5 job. Expect to work more than 40 hours a week without overtime pay.

prospects: Every state has a wildlife rescue team, but staffs are small. However, because of the long hours there is a lot of job burnout, thus frequent turnover. You won't find "stranding coordinator" jobs listed in the newspaper want ads. You'll have to seek out the operations yourself. The best bet is to step into a rescue team operation by volunteering or through an internship.

qualifications: You should have at least a bachelor's degree in biology and a veterinarian technician certification, which many community colleges offer through two-year programs.

characteristics: The job requires tremendous patience and a gentle touch with which to gain the trust of wounded, abandoned, or wayward coastal and marine-life animals.

Kim Durham *is a stranding coordinator on Long Island, New York.*

How did you get the job?

Kim Durham says that when she was a child she wanted to be a veterinarian. She was especially interested in marine life. "I was one of those little girls who loved *Flipper,*" she says. After graduating from college with a biology degree, Kim didn't know what she wanted to do, so she headed back home to Long Island and began doing volunteer work at Okeanos Ocean Research Foundation. She volunteered for three years before getting hired onto the staff.

What do you do all day?

"This morning I was beeped at 7 o'clock to check on a stranded seal," Kim says. "Depending on the situation, I might take along another staff member or volunteer to help with the rescue."

CONTACT YOUR STATE DEPARTMENT OF ENVIRONMENTAL CONSERVATION FOR INFORMATION ABOUT WILDLIFE RESCUE ORGANIZATIONS IN YOUR AREA.
ONE HINT IS TO APPROACH AN AGENCY DURING A BUSY SEASON WHEN IT CAN USE EXTRA HELP. FOR EXAMPLE, IN THE NORTHEAST MOST SEAL STRANDINGS OCCUR DURING THE WINTER MONTHS.

It's a cold winter morning on the Long Island shore and this time Kim is alone on the rescue. First she evaluates the animal's condition. Next she throws a sheet over the seal's head and manually lifts it into a transport cage. The job certainly has its hazards. "The seal bit right through my pants ... luckily it didn't get my skin," she says. Kim then brings the seal to the Okeanos facility where she draws blood samples and prepares the animal to be checked by a veterinarian.

Mornings are the busiest time of day at Okeanos. A typical day begins with feeding the animals and the unglamorous job of cleaning out cages and tanks. Kim also administers medications, takes photographs, and tags animals.

Recently, there were eight seals and three sea turtles in her care. Some of the animals were sick from natural causes, while others were victims of human cruelty. Two of the seals had been brutally slashed. There's also the unexpected. This particular afternoon a fisherman dropped off 1,200 pounds of fish as a donation. So suddenly Kim found herself completely surrounded by fish, and she was off to help pack 1,200 pounds of fish into plastic freezer bags.

Kim always wears a pager and can be beeped any time of day or night to answer a call that there's a stranded or injured seal, dolphin, or disoriented whale washed ashore. "The toughest part of the job is that on any particular day you may have worked 14 hours straight and then, in the 15th hour, be called out on a stranding to save an animal. I love that every day there's something different about the job," she says. Literally the biggest and most satisfying rescue of her career so far was a team effort to save a sick and disoriented pilot whale. The gentle whale was successfully rehabilitated and returned to the sea.

Where do you see this job leading you?

Kim has decided to return to school to pursue her childhood dream of becoming a veterinarian specializing in marine animals.

2. Wildlife Rescuer/Rehabilitator

description: The job of a wildlife rehabilitator is to rescue injured or abandoned land animals, such as mountain lions, birds, wolves, bears, and deer. The goal is to nurse these animals back to health so that they can return to the wild. If that's not possible—for example, with a bird whose wings have been clipped—the bird or animal is permanently housed at a wildlife refuge. Rescuing animals is not a 9-to-5 job. These saviors are on call 24 hours a day, including weekends. When not out on calls, the wildlife rescuer feeds and cares for the animals. That includes the dirty work such as cleaning out cages. The rehabilitators also administer medical care.

salary: People who do this work don't do it for the money. These jobs tend to be with nonprofit organizations. Salaries are notoriously low. Starting pay is approximately $6 an hour.

prospects: Every state has at least one rehabilitation center. Staff positions depend on local income and fund-raising. Budgets are small, so staff jobs may be hard to come by. The flip side of this is that the work and hours are so grueling that there tends to be a lot of turnover as a result of burnout.

qualifications: A bachelor's degree in biology or veterinary science is suggested. Some colleges offer a two-year veterinarian technician certification.

characteristics: Stamina, patience, fortitude, and a gentle manner in caring for injured and mistreated animals are the characteristics one needs for this line of very demanding work. You need these characteristics not only to deal with the animals and the grueling realities of the job, but also to deal with the public—in this position you're likely to spend as much time relating to people as to animals. Many of the people who call on you are cordial, but occasionally you're on the receiving end of some rotten attitudes. "Sometimes someone will call screaming, 'Come get this animal now or I'll shoot it,'" says one animal rehabilitator.

Tim Ajax *is an assistant director of wildlife rehabilitation and rescue in Boerne, Texas.*

How did you get the job?

Tim Ajax doesn't have a science background. He's a musician and was working as a camera technician, sitting on a bench 10 hours a day fixing expensive cameras, when he began volunteering at the Wildlife Rehabilitation and Rescue Refuge. He found himself so immersed in the work that he was spending every minute of his free time there caring for animals. Finally, he was offered a job. Today his responsibilities are to oversee and manage the 21-acre refuge, which is not open to the public. Depending on the time of year, the refuge has 150 permanent resident animals and as many as 800 orphaned and injured animals. "Right now we have 13 mountain lions, 4 jaguars, 3 wolves, 3 black bears, and 20 bobcats, plus hundreds of animals like raccoons, possum, monkeys, and birds."

What do you do all day?

"When I first get in I make the rounds of all the new animals. Some 10 to 30 new animals come in each day, and some days it may be as many as 80," Tim says. "Some of the animals were found injured, some found orphaned like the four baby raccoons we found whose mother had been shot, or the possum that was run over and its babies flew off her back." Tim says many more animals brought into the refuge are victims of the exotic pet trade. "They are from animal owners who thought it would be cute to own a monkey, then changed their minds." In making his rounds, Tim checks the health of each animal and to see if medications are working. If needed, he'll adjust feeding schedules. Later he may answer the phones while other staffers catch up on their work. "Suddenly, if one animal takes a turn for the worse and may die, you drop everything to treat it." The day may be filled with all kinds of calls—a deer hit by a car, raccoons in the ceiling, bats on the front porch, or the police might call to report that someone has spotted a rattlesnake in their front yard. "You just learn by experience, making it up as you go," Tim says.

Tim offers a warning. "Everyone who comes through the door has a romantic view of the job of helping wildlife. Even when I explain how tough the work really is, they still don't understand how much so until they get into it." It's not just a matter of the hours and dirty work, he warns, but also the sometimes difficult moral and ethical decisions rescuers must make. "There are times you have to decide to destroy an animal rather than let it suffer. You have to learn how to step back."

Tim has never been seriously bitten or scratched. However, he's faced some scary situations. The scariest was when he had to step into a tiny room to tranquilize a mountain lion locked up in an abandoned house.

Where do you see this job leading you?

Tim wants to become a veterinarian. Because he feels that there is not enough research on wildlife medical treatment, he wants to try veterinary medicine and develop an expertise in wildlife medicine. He already finds himself forced to experiment and create medicinal formulas on the job.

> **ASK THE LOCAL WILDLIFE REFUGE WHEN THEIR BUSIEST SEASON IS, AND BE PREPARED TO JUMP IN THEN TO HELP. VOLUNTEER. ALMOST EVERY STAFF MEMBER OF THESE ORGANIZATIONS HAS VOLUNTEERED THERE. YOU MIGHT ALSO TRY GETTING AN INTERNSHIP.**

3. Field Biologist–Birds

description: An ornithologist specializes in the study of birds. This specialist examines the behavior of birds, which involves everything from their feeding and nesting to their responses to the effects of the alteration and destruction of their habitats. Don't think you'll spend all of your time in the great outdoors watching birds. Two-thirds of the year is spent indoors analyzing data and writing reports and journal articles. As a novice, you'll most likely answer to a senior scientist heading a project or department.

salary: Entry-level field biologists/research assistants earn salaries starting at $20,000 a year.

prospects: There are permanent staff research jobs with observatories and with nonprofit environmental protection and conservation agencies. Many projects and jobs are funded only for a particular length of time. The best way to get field experience is through seasonal jobs. Projects often hire teams of temporary technicians for fieldwork. The toughest part is finding out about job openings, which requires constant networking.

qualifications: A degree in the sciences, be it in biology, chemistry, ecology, or a related field, is usually required.

characteristics: Patience and flexibility are essential, as the job can require living and working in a group situation over a period of time as the team researches a particular project.

Peter McKinley *is a field biologist specializing in the study of birds.*

How did you get the job?

With a master's degree in biology, Peter McKinley wanted to study the effects of the timber industry on birds in the Northeast. A college adviser gave him a contact's name. But that was just a start. Peter says it was his well-written personal letter indicating his committed interest in the observatory that helped his resume stand out from the crowd. Instead of sending out a hundred resumes, he targeted this particular observatory and persisted until he got the job.

What do you do all day?

Three months out of the year, Peter is out in the field with a crew of eight technicians that he's hired. They spend 10-hour days, six days a week, counting and tracking birds in a study that covers a 50-mile radius.

> THERE IS PLENTY OF SEASONAL WORK TO BE HAD. ALTHOUGH THE PAY IS OFTEN VERY LOW, THIS IS A GREAT WAY TO GET THE RESEARCH EXPERIENCE YOU'LL NEED TO GET YOUR FIELD BIOLOGY CAREER STARTED.

During that time they may track 80 different species of birds, mapping their territories and following a scientific checklist of observations.

"We usually rent a house where all the crew members live. You've got to have the people skills to effectively manage the team, live all together, and efficiently complete the research," Peter says. The rest of the year is spent writing grant proposals, research papers, and articles for journals. "I also work on the fund-raising," he says. "Many students are disillusioned [when they find out] what a field biologist does. They think we spend all of our time roaming the outdoors in a Range Rover, when actually most of our time is spent inside analyzing data," Peter says. The observatory scientists use the research data to advise timber companies how to mitigate harm to the birds' habitat. For example, the scientists may suggest that the companies adjust the size or area of a harvest. Peter answers to a senior scientist in charge of the project.

Where do you see this job leading you?

"I'm going back to school to get a doctor's degree in ecology," Peter says. Without it, he is restricted to working as an assistant on research projects. He says a doctorate is necessary to head up his own projects. "You need the analytical and computer skills that a Ph.D. requires. It's also difficult to win grant money without a Ph.D."

4. Species Coordinator

description: A species coordinator manages an entire endangered animal species. Species coordinators work under the auspices of the American Association of Zoological Parks and Aquariums. Some years ago, the association created Species Survival Programs (SSPs) to help preserve and protect endangered animals such as gorillas, rhinos, and tigers, as well as different bird species, reptiles, and amphibians. There are approximately 70 SSPs and a coordinator for each one. Using a computer program, coordinators track animal populations and their demographics. They make breeding recommendations based on the data. They direct which animals should be paired and bred. These things are done to keep the endangered species as diverse as possible. Most SSP coordinators work for zoos and do the SSP work as part of their job as zookeepers or zoo directors or veterinarians.

salary: Most SSP coordinators do this work along with other duties as zookeepers. Starting salaries for zookeepers can be as low as $15,000, while zoo directors can earn $70,000 to $80,000 a year.

prospects: This is a very specialized and therefore somewhat limited field. While there is only one full-time species coordinator in the United States, the American Association of Zoological Parks and Aquariums is adding new SSP programs every day. Not all SSP programs involve zoos and captive animals—the U.S. Fish and Wildlife Service oversees some endangered wild animals such as the Florida panther and the California condor.

qualifications: More and more zoos require a bachelor's degree for applicants wanting to be hired as a zookeeper. A wildlife biology, zoology, or animal behavior degree is the way to go. Computer skills are a must.

characteristics: You must be a stickler for detail, staying in touch with zoos worldwide to keep careful track of particular animal populations. Of course, you have to love animals. Patience is also a must, as animals have their own personalities; a computer data analysis may recommend that particular animals should mate and breed, but often the animals don't agree.

Lori Perkins *is a conservation biologist/species coordinator for a zoo.*

How did you get the job?

Even with bachelor's and master's degrees in psychology, Lori Perkins had to start at the bottom and work her way up. With the help of one of her professors, she got a job at the zoo doing clerical work as a registrar. Then the opportunity came for a position as a Species Survival Program coordinator for orangutans.

What do you do all day?

Lori's top responsibility is keeping a "stud book" on orangutans—a book that contains gene information as well as the lineage of each of the ape species. She keeps track of all the orangutans in zoos world-

FIND OUT WHO'S DOING SSP WORK AND IF ANY NEW PROGRAMS WILL BE ADDED AT THE ZOO. GET IN TOUCH WITH THE AMERICAN ASSOCIATION OF ZOOLOGICAL PARKS AND AQUARIUMS TO STAY ABREAST OF NEWS ABOUT NEW SPECIES SURVIVAL PROGRAMS.

wide and of as many in the wild as is possible, but she manages the breeding of orangutans only in North America. Lori's job is mostly data management and not research. She is on the phone every day and is in touch with at least four or five zoos, checking on the orangutans. She knows each one by name and knows their personalities . . .

all 253 of them in the United States. Births and deaths of the animals can change the genetic profile of the entire population, so she is constantly running computer analyses.

Recently, Lori was faced with the dilemma of finding the perfect mate for an orangutan named Otis. Otis, she says, carries a rare gene and she must find a female with the appropriate genetic match. But the problem doesn't end there. Otis isn't exactly romantic; in fact, he has a reputation for beating up on females. Lori works with a management committee that oversees each of the endangered species programs so that it's not just one person's ideas and analysis determining what's happening with the population.

As part of her job, she has traveled to Indonesia to help that country develop a conser-

vation plan for orangutans in the wild. When Lori entered the job she knew nothing about genetics and not much about orangutans. She was forced to take it upon herself to take courses and do research.

Where do you see this job leading you?

Lori wants to expand the job by doing more in the area of education. "We need to justify the existence of these animals in captivity. We need to find the best ways to use orangutans to educate the visiting public about their conservation needs." Lori says down the road she wants to get a Ph.D. and, as a new challenge, may switch from orangutans to running a program for another endangered species.

5. Exotic Animal Nutritionist–Zoo

description: Feeding zoo animals is much more than just cutting up carrots and apples. The daily preparation of meals for hundreds of different exotic animals and birds is a daunting responsibility. The animal nutritionist is responsible for determining a healthy, appetizing diet for each animal. The nutritionist must also order food, track inventories, and make sure food is properly delivered; you don't want raw meat to be sitting out in the hot sun all day. The responsibility of an animal nutritionist is to ensure that the animals get healthy, nutritious meals and not merely culinary delights. However, like regular chefs, the zoo nutritionist must please his or her clientele by coming up with new concoctions for those animals with finicky palates. Just like humans, some animals get bored with the same dishes. To come up with new nutritious entrees for a yak requires research and experimentation.

salary: Zoos aren't known for paying high salaries. The starting salary for this job may be $15,000. But people who work at zoos don't do it for the money; they do it because they love it.

prospects: Zoos across the country are just waking up to the idea that they need exotic animal nutritionists. This is such a new career that you're not going to find many zoos with exotic animal nutritionists on staff. The career was not formally recognized by the American Association of Zoological Parks and Aquariums until 1975. Before that, zoos just relied on each other to determine what to feed a particular animal; they checked to see what was working at other zoos. Since zoos are exploring this new specialization, the prospects are good for anyone who can demonstrate a strong animal nutrition background.

qualifications: At this writing, there are just a few colleges or universities that teach nutrition for exotic animals. Veterinary schools teach nutrition courses for domestic animals. However, you should have a degree in animal nutrition, zoology, animal behavior, or a related field. One should also have a strong background in plant sciences or botany.

characteristics: Resourcefulness is at the top of the list. No one's teaching what to feed zebras and giraffes, so the job requires someone with initiative to research and learn on his or her own. One also needs to be a problem solver and to seek answers to the new questions that continually surface.

Gloria Hamor *is an animal nutritionist at Zoo Atlanta in Atlanta, Georgia.*

How did you get the job?

After Gloria Hamor decided against going to veterinarian school, a college adviser suggested she study animal nutrition. But her advisers were taken aback when Gloria took the idea one step further and asked to study exotic animal nutrition. There weren't any such classes. Universities were teaching nutrition only for farm animals such as cows, sheep, and pigs. Her adviser connected her with the local zoo, where she did her thesis. She joined the American Association of Zoological Parks and Aquariums and saw the Zoo Atlanta position advertised in the organization's monthly newsletter. At the time (1984), most zoos did not have nutritionists on staff. Now zoos are discovering the benefits of someone with an animal nutritionist background.

> THE BEST ADVICE IS TO START BY VOLUNTEERING AT A ZOO. JOBS AT ZOOS ARE VERY COMPETITIVE. MOST OF THE PEOPLE HIRED COME UP THROUGH THE RANKS AS VOLUNTEERS. EVEN THOSE WITH ANIMAL SCIENCE DEGREES MAY START BY SELLING TICKETS AT THE GATES. JOIN THE AMERICAN ASSOCIATION OF ZOOLOGICAL PARKS AND AQUARIUMS. THE ORGANIZATION'S NEWSLETTER OFFERS JOB LISTINGS.

What do you do all day?

Walk into the giant kitchen with its gleaming stainless steel preparation tables and gigantic freezers, and you'd think you were in the kitchen of a popular restaurant. But one look at the box of frozen crickets in the freezer or the barrel of dry feed in the corner, and you realize this kitchen isn't for human consumption. On any given day Gloria is in charge of preparing 110 menus. She checks inventories, orders food, updates vendor lists, and makes price comparisons. Today she's literally ordering food by the ton, as she orders dry feed. She double-checks diets so that the correct meals get to the correct animals on the correct days. A typical day may find her checking with the zoo veterinarian staff to see if a new animal is coming in. If one is being transferred to another zoo, she'll write up a form explaining the animal's diet. Then there's the unexpected: Gloria might have to run out to the grocery store to pick up a jar of jelly because one animal is taking medication and will swallow it only with grape jelly. What you also won't find in a restaurant kitchen is the "live food" room. In a strange twist, Gloria also has to feed and maintain the animals, such as rabbits, rats, mice, and chickens, that will eventually serve as food for other animals.

Where do you see this job leading you?

Gloria loves the continuing challenge the job and the animals offer. She hopes to see more people enter this field and research and create educational curricula. In the meantime she works with the American Association of Zoological Parks and Aquariums nutritional advisory group, created so that nutritionists from other zoos can compare notes.

6. Zookeeper

description: A zookeeper is assigned to the daily observation and maintenance of the overall well-being of a specific group of animals within the zoo. While a zookeeper's work is fascinating, intriguing, and rewarding, it also can be a dirty job. A zookeeper is responsible for feeding the animals in his or her care and cleaning out their cages. A zookeeper's duties, however, go beyond the daily care of the animals. The keeper often develops an intuitive sense, a rapport with the caged creatures. He or she looks for behavioral changes. The simple fact that an animal is not eating well may indicate that the animal is sick and must be treated. Over time the zookeeper's dedication, love, and care for the animals becomes instinctive, almost like a mother's love.

salary: Nobody does this work for the money. Salary depends on the type of zoo facility. Privately operated zoos tend to pay less than city- or county-run zoos, where the benefits are on a par with most city jobs and the pay is reasonable. At city- or county-operated zoos, keepers start near $20,000 a year.

prospects: Jobs are few and far between. The number of zookeeper jobs is limited and applicants are many.

qualifications: Twenty years ago college degrees were not a prerequisite, but competition for these positions has changed that. Zoology and/or biology degrees are recommended. Some colleges even offer degrees in zookeeping. A shorter educational route is a veterinary technician degree, which can be earned in less than two years.

characteristics: Dedication is probably the most important quality one needs to be a zookeeper, but the work also requires you to be as comfortable handling a venomous snake as a cuddly panda. Keen observation skills are critical, as is the ability to make quick assessments of animal behavior.

Cindy Bickel *is a zookeeper.*

How did you get the job?

While she was in high school Cindy Bickel began working at the zoo during the summer months, often in the children's petting area. She started college, but quit when she got a full-time job at the zoo. That was 25 years ago. Back then, she says, zookeeper jobs weren't that desirable and not many women were in the field. Now degrees are required and it's a very popular and competitive job. She says there are an average 200 applicants for each job opening. As a full-time staff employee, she started as a "relief keeper" assigned to various sections of the zoo. Eventually she gained a permanent position working in the hospital.

What do you do all day?

In the hospital, Cindy is charged with providing basic medical care to ailing animals, attending newborns, and checking the health of those being transferred into the zoo. The day starts out with her checking each of the hospitalized animals. Then it's time to begin administering medical treatments: everything from giving antibiotics to birds or putting ointment on a frog to giving an injection to a tiger. On any given day there may be as few as 6 patients or as many as 30. She works with two full-time veterinarians, a medical technologist, another zookeeper, and at least two trained volunteers. Besides the medical care and nurturing she gives the animals, Cindy also sweeps and cleans the animal cages. Although there are scheduled procedures such as the primates' yearly physical exams and dental checkups, the days can be unpredictable. An emergency might call for Cindy to assist the veterinarian in sedating an ailing lion.

Cindy has suffered her share of bites and scratches. "Oh yes, it's very dangerous. You have to be extremely careful, especially if you have a large carnivore like a leopard or tiger. You're prone to being bitten, kicked, or gored," she says. Cindy once had a close call when she was nearly entangled by a python. "I was giving him a bath ... He started squirming out of my hand and when I went to grab him, he just barely got his mouth on the side of my hand and wrapped around both of my hands," she says.

There is no question that Cindy's dedication is unsurpassed. Last year she even made it onto the network news when twin baby polar bears were born at the zoo. Cindy, another zookeeper, and the zoo veterinarian became surrogate mothers for the cubs, who had been rejected by their mother. Just like human twins, the baby polar bears required round-the-clock attention. Every night for three months she took the bears home with her. "It was just like having twin babies; if one was sleeping, the other was awake and fussy." Now the bears, named Klondike and Snow, each weigh more than 200 pounds and live in their own enclosures.

Where do you see this job leading you?

Cindy loves her job and wants to stay where she is. She says that because the job isn't very glamorous, many zookeepers move on to other work. Many of them, she says, go back to school for graduate degrees in specialties such as animal behavior, nutrition, and veterinary medicine.

> VOLUNTEERING AT THE ZOO IS THE BEST STRATEGY FOR GETTING YOUR FOOT IN THE DOOR. YOU SHOULD ALSO BE WILLING TO START IN OTHER JOB CAPACITIES WITHIN THE ZOO. EVEN THOSE WITH COLLEGE SCIENCE DEGREES START OUT BY TAKING TICKETS AT THE ZOO'S GATES.

7. Aquarium Keeper

description: An aquarist is a zookeeper who specializes in marine animals. Every large aquarium employs people to care for the creatures on display, but the folks who do this type of work need to have an extra measure of dedication. Not only does the job entail all the work of feeding, caring, observing, and evaluating that comes with a conventional zoo, there is the pervasive presence of the water environment to contend with and preserve. Most beginning aquarium keepers will come to a facility directly out of college, often after volunteering. People in this job are united by an affection for animals and a capacity for hard, sometimes backbreaking, work.

salary: Salary depends on geographic location. Some small aquariums have advertised these positions for as little as $5.50 an hour. An aquarium in a large city might start you out at $25,000 a year.

prospects: Established aquariums have a low turnover rate; the New York City Aquarium, for instance, hires one or two aquarium keepers a year. However, a number of new aquariums have opened in recent years in Florida, New Jersey, Louisiana, California, and New England.

qualifications: A degree in one of the sciences is preferred. Scuba certification is almost always a must. Usually an aquarium keeper will have had some experience with marine animals from volunteer or school-related work. You should be physically fit; carrying heavy bags of fish food and diving into a cold tank in the dead of winter require a good degree of stamina.

characteristics: Perhaps the most important qualities an aquarium keeper can exhibit are a keen sense of observation and the quick application of knowledge about the animals in his or her care. The animals are, after all, in a water environment whose chemistry and temperatures are critical to their survival. Since the pay is relatively modest and the work sometimes physically demanding, a strong sense of dedication goes a long way toward making this job choice successful. It is important to be able to think critically and to be able to assess the condition of an animal based on observation.

Hans Walters *is an aquarium keeper/supervisor at the Bronx Zoo in New York City.*

How did you get the job?

Hans Walters had been working with birds at the Bronx Zoo when he saw some aquarium jobs advertised in trade organization newsletters. He had a degree in marine biology and a scuba certification, so he decided to apply. He is now in charge of the area that includes sharks, penguins, and walruses.

What do you do all day?

Hans says he does everything a zookeeper does (and that means feeding, tending, and caring for the animals) except that the task of cleaning the animals' enclosures is complicated by the fact that they're filled with water.

Many of the tanks cannot be emptied for cleaning, so that means putting on scuba gear and diving in, sometimes several times a week, even in the dead of winter—perhaps the worst experience you will ever have. Even though Antarctic dry suits are used in winter, it is still a bone-chilling job. Yes, Hans gets in the tanks with sharks, and he admits the first time was unnerving. "The first time I got in a tank with them my eyes were as big as saucers," he says.

> **GET A DEGREE IN AN APPROPRIATE SCIENCE AREA. VOLUNTEER. GET SCUBA-CERTIFIED AND PRACTICE.**

The first thing Hans does in the morning is to make sure the animals are content. His responsibilities also include checking the pumps and the water temperature and noting whether or not the glass is covered with algae. Keepers of the marine mammals also spend time "bucketing up" the day's food, which means defrosting the frozen fish and dividing it among the various animals. Part of the job of aquarium keeper includes being an educator. At scheduled feeding times the keepers often interact with visitors, offering demonstrations and information about the animals. Much of the day is spent cleaning up and making sure the place is presentable. However, the job is much more than feeding and keeping tanks clean. It often requires troubleshooting. For example, you may walk in and notice the tank is cloudy, so you have to try to figure out why and try to fix it, all the time watching the condition of the animals for advance warning signs of possible health problems.

Be prepared to work long hours and weekends. It's a 365-days-a-year job.

Where do you see this job leading you?

From a keeper's position the natural move up is to become a supervisor managing the work of others with less experience. From there, the career path may lead to an assistant curator's job or some other function in the aquarium's administration.

8. Marine Mammal Trainer

description: When most people think of a marine mammal trainer's job, they envision bottlenosed dolphins cascading through water on their tails or leaping through hoops. But teaching tricks or behaviors to dolphins is only a small part of the many tasks that a marine mammal trainer will undertake throughout his or her career. Marine mammal trainers work with numerous marine mammals, such as walruses, sea lions, and beluga whales. Aside from teaching them behaviors, trainers are responsible for these animals' general well-being. This means making sure that the animals' pools are clean and that they are getting the right amount of food as well as the right amount of activity. Trainers can become an integral part of the lives of the marine mammals they care for, and in return the trainers are given a glimpse into the mammals' world. Marine mammal trainers discover much about these animals' thought processes as the animals are challenged with more and more difficult chores. When a new behavior is mastered, the trainers share in the animals' excitement.

salary: Someone just starting out in the field will more than likely be working in a seasonal or part-time capacity and earning minimum wage. Full-time positions for trainers pay around $17,000 to $20,000 a year. Senior mammal trainers can make $30,000 a year and higher.

prospects: Because the pay in this field is not very high, those who work in it are generally dedicated to what they do. Consequently the turnover rate is low and jobs do not become available very often.

qualifications: Although some marine mammal trainers today do not hold a bachelor's degree, it is becoming increasingly difficult to break into the field without one. A bachelor's of science in biology, or better yet marine biology, will provide any aspiring marine mammal trainer with the necessary background for the field.

characteristics: While all marine mammal trainers must have a love for the animals, this affection can take them only so far. It is also important that trainers have patience. Caring for and teaching animals is similar to doing the same with a child. If a trainer becomes frustrated easily or tends to enter into ego struggles with the animals, he or she won't be successful.

Martha Hiatt *is a senior marine mammal trainer.*

How did you get the job?

Growing up, Martha Hiatt was interested in marine mammals, especially dolphins and whales. She always knew that she would end up working in some capacity with marine animals, but she didn't think it would be as a trainer. "I wasn't at all interested in training—I thought it was much more of a power struggle type of thing between the animal and the trainers and I had a very negative impression of it," Martha says. But Martha's strong convictions about this profession began to change when she was 23 and a volunteer at the New York Aquarium. She was required to conduct general cleaning duties and prepare food for the marine animals. She was also permitted to sit in on some of the training sessions. "The first time I saw a dolphin training session, I knew it was what I wanted to do," Martha says, explaining that she saw the dolphin and trainer engaging in an organized game rather than any type of power struggle. Martha says she watched as the trainer at the aquarium took a dolphin through a series of training

steps. He seemingly wasn't making much progress until "suddenly it hit the animal what the trainer wanted him to do, and the excitement that the animal displayed is what got me. I saw right then and there how stimulated the animal was."

After she had spent six months as a volunteer for the training department, a part-time assistant trainer position opened up, and Martha took it. A year later, she was given a full-time training position, and four years later she was promoted to the position of senior trainer.

What do you do all day?

Martha's entire workday as a senior trainer is centered around the animals. In an average day, she might conduct 20 to 30 different training sessions in conjunction with feedings or shows. She starts her day at 8 A.M. by

thawing out the first portion of the 500 pounds of fish she will throw to the animals in a day. This is more than just breakfast for the animals; it is the first training session of the day. Though the animals are sometimes rewarded for performing, they are always fed regardless of whether they perform or not. Some of the things Martha asks them to do will be part of their public performance later that day, but she also teaches the animals behaviors that can come in handy when the doctor comes for a visit. For example, she may teach a dolphin to put his tail up on the poolside and hold it still so blood can be taken or a shot can be administered. During the morning feeding/training session, she pays close attention to the animals and their surroundings. If any of the animals are in need of medical attention, she will

arrange for it. The afternoon will usually consist of a performance where the animal can show off its behaviors in front of an audience. Martha says she also considers this to be a training session, because during the shows she has the animals repeat the tricks they have learned in training. She also takes the opportunity during the show to raise the public's awareness of these animals' needs. "By educating the public, I can try to help develop an affinity between the public and the animals. The affinity that the public has with the animals gives them the impetus to take care of them."

Even when the show is over, often it's still not time to go home. Martha stays on call 24 hours a day. If any emergency situation arises, she is called in to take care of the animals.

Where do you see this job leading you?

Martha definitely wants to continue working with marine animals. The next step for her would be to become the director of training. After that, she could move on to become the park's curator.

9. Wildlife Biologist

description: Wildlife biologists survey and monitor wildlife population and activity in a designated area. This type of work is usually done for a national park, forest, refuge, or other federal lands and might involve determining the effects of recreational activities at the park on wildlife. On the other hand, a wildlife biologist working for the state might work to enhance a certain population of game animal. Still other wildlife biologists work for private nonprofit organizations in which wildlife conservation is generally the main focus of their job. In this sort of a work setting, a wildlife biologist might gather data to be used to raise public awareness of a specific animal's plight, such as that of the bat or the spotted owl. This job may also entail a lot of speaking to the public—private landowners as well as state and federal officials.

salary: The starting salary for a government wildlife biologist is about $20,000 per year, and the pay increases according to the amount of experience. For example, after two years of experience the pay can go up to about $25,000; with promotions, this can increase to about $40,000. In a nonprofit agency, the pay can start as low as $19,000, but it can climb as high as $30,000 to $40,000.

prospects: The market for wildlife biologists is very competitive—there are not that many jobs out there. Oftentimes people just starting out will settle for seasonal work or a job in a very rural location just to break into the field.

qualifications: A bachelor's degree is required for this kind of work. It is best to focus your studies in the sciences, preferably biology. Because the field of wildlife biology is so competitive, it wouldn't hurt to put yourself ahead of the bunch by obtaining your master's or doctorate. Getting experience in the field by volunteering at a national park or for a graduate student's study is another good way to make yourself stand out as a candidate for employment.

characteristics: Because most of the work of a wildlife biologist is done outdoors, people in this field must like to be outside. They usually share a love of wild animals as well, and are genuinely concerned about the overall well-being of wildlife.

Dan Taylor *is a wildlife biologist specializing in bats.*

How did you get the job?

While Dan Taylor was working to get his bachelor's degree in wildlife biology from the University of Montana, he spent his summers getting experience in the field. For two seasons he worked as a wildlife technician for the National Forest Service in Kootenai National Forest. Dan also gained experience during the winters by working on an indoor data compilation of wildlife studies as part of a college work-study job. After graduation, he went into the Peace Corps for two years, where his college experiences allowed him to secure a position as the director of the Cockscomb Basin Jaguar Preserve in Belize. After the Peace Corps, Dan was able to gain full-time employment as a district biologist in the Mount Hood National Forest in Oregon. He then transferred to a similar position at Prescott

> THE BEST RESOURCE FOR FINDING JOBS IN THIS FIELD IS THE NATIONAL WILDLIFE FEDERATION'S CONSERVATION DIRECTORY, WHICH LISTS ALL WILDLIFE ORGANIZATIONS IN THE UNITED STATES. (FOR ADDITIONAL INFORMATION, SEE RESOURCES, PAGE 212.)

National Forest in Arizona, and now works for Bat Conservation International in Austin, Texas, as the director of the North American Bats and Mines project, which is geared toward raising awareness of the importance of abandoned mines as bat habitats.

What do you do all day?

Over the last 100 years bats in North America have been forced out of their traditional roosts due to recreational activities and the destruction of caves and tree hollows. Many bats have found alternative shelter in

abandoned mines. However, mining companies and government officials have methodically closed off these abandoned mines, leaving bats trapped inside, where they die. For the most part, Dan spends his time speaking to government and mining company officials to raise awareness of the problems facing bats in North America. He urges these officials to check for bats in abandoned mines before sealing them off and to install a bat-friendly gate when shutting the mine down rather than sealing it off completely.

Dan also spends a lot of time out of the office conducting workshops in which he shows land management officials how to determine whether bats are roosting in a mine. During these workshops, he demonstrates how to close mines in a fashion that protects both bats and humans. With all of these workshops and speaking engagements, Dan spends most of his time out of the office. He uses any office time he has to set up other engagements.

Where do you see this job leading you?

Dan says he doesn't plan to stay in this particular position forever; however, he hasn't yet decided on his next move. From here he could go into independent consulting, a higher-level position for another nonprofit organization, or a higher-level position with the federal government.

10. Wildlife Biologist

(FOREST PRODUCTS COMPANY)

description: This is a new one. Logging companies that harvest trees for paper and lumber are beginning to hire wildlife biologists to help them meet and shape federal regulations. It's partly a matter of "if you can't beat 'em, join 'em" and partly "let's do what we can and talk some sense into these people." In this setting, a wildlife biologist will look over the shoulders of federal and state wildlife officials, and help the forest products company do its job while avoiding the confrontations of the past. By advising the company on how it can minimally change its procedures to preserve wildlife habitats, the forest wildlife biologist can reduce the cost of compliance and help generate goodwill and public relations benefits for the firm. The process, it's hoped, will ensure that a lot of animals still have homes after the loggers finish doing their job.

salary: Salaries average $25,000 to $35,000 a year.

prospects: Companies are just beginning to get into this area. In many ways this is the traditional wildlife biologist's job, but with a private industry rather than a government regulatory slant. You should look for this position at firms that own private forest land subject to government environmental regulation, such as paper or forest products companies.

qualifications: A master's degree in environmental studies is required. A forest wildlife biologist should also have some administrative skill and knowledge of the regulatory process. Familiarity with the specifics of the appropriate business is a plus, but on-the-job training is usually available. It is extremely helpful to be computer literate, especially in GIS (Geographical Information System), which is used to create map models. You need not only to be knowledgeable about wildlife, but also to have a good, basic scientific background.

characteristics: Aside from the science, this job is primarily about communication. You have to be comfortable talking with both loggers and top company executives. Given the passions that surround many of the issues involving loggers and wildlife, it won't do to be abrasive or arrogant about your ideas. In this job, you are working within the system that, in some sense, is creating the problems you want to address.

Jessica Eskow *is a wildlife biologist with a forest products company.*

How did you get the job?

How does a young woman from New York City find herself working in the middle of an immense forest? Jessica Eskow became interested in wildlife as a child growing up near the Bronx Zoo (and yes, there are some trees in New York City). She got a bachelor's degree in biology and a master's in forest science. While in college she did summer work as an assistant wildlife manager for the Audubon Society and the Sierra Club. State jobs are pretty bureaucratic, Jessica says. She wanted the flexibility to carry out her own ideas, and she is philosophically opposed to excessive emphasis on government regulation as a means of solving problems, so she sought out major timber companies. "I got a list of all the companies that were members of the AFPA, the American Forest Paper Association," she

explains. "They have a wildlife committee, so I faxed each committee member a letter asking about a job in my field. Then I faxed them a resume, and called them up and kept harassing them for several months." Jessica interviewed for jobs in Maine, Connecticut, South Carolina, and Georgia before finally landing her job in Washington state.

What do you do all day?

Jessica has started several research projects, while at the same time she inspects harvest areas, and advises the company about the possible presence of endangered species and how to deal with them. "We have a number of spotted owls on our land," she tells us. "That's a priority. The other day, though, a state wildlife biologist found a Van Dyke salamander on one of the units they're about to harvest, so I had to check that out as well. We have to see what kind of habitat they have and how we can maintain it." She

> TO GET A WELL-PAYING JOB WITH A FOREST PRODUCTS COMPANY, YOU'LL NEED TO ATTEND A FORESTRY SCHOOL. BUT ALSO GET AS MUCH EXPERIENCE WORKING WITH WILDLIFE MANAGEMENT AS YOU CAN, WHETHER IN A FOREST, A PARK, OR A ZOO.

also keeps watch over the health and habitat of larger game animals such as elk. "I'm responsible for the populations' being at good levels," she says. This summer she's involved in collaring and radio-tracking elk to see how the logging affects their movement.

Jessica makes it a point to learn every aspect of the business. On this particular day she was on her way to the log yard to see what is done there. "I need to know the operation from start to finish and the limitations with which I have to operate." The reason Jessica loves the job is that the career

field is so new that "they'll let you try almost anything within reason." She says this job fits with a commitment to protecting wildlife. "My philosophy about the whole habitat issue and providing wildlife with conservation opportunities lies with industry and not with regulation from government." It's not a 9-to-5 job. Jessica says the work hours can be "crazy." She's at work by 7:30 in the morning, and if she's doing fieldwork she often stays beyond 6 o'clock in the evening.

Where do you see this job leading you?

Jessica wants to do the type of research that will help develop new strategies for preserving wildlife in industrial forests. For instance, she points out that dead trees, or "snags," used to be taken out of logging areas as a hazard to workers. But research helped the companies reevaluate that practice. The snags, it turns out, are valuable habitats for a variety of animals, including woodpeckers.

11. Wildlife Biologist

(PRIVATE CONSULTING FIRM)

description: A wildlife biologist examines and evaluates the relative health of a species within an area. By visiting the habitat, making observations, and applying his or her knowledge, the biologist can develop an opinion about what is happening to an animal population or what is likely to happen if the habitat is altered. For a private company trying to comply with environmental laws and regulations, the wildlife biologist's work is essential to ensure that new construction or operational change can be accomplished without violating those laws or regulations.

salary: Salaries at private firms start at around $21,000 a year. As might be expected, salaries with nonprofit groups may not be as high.

prospects: Consulting firms are looking for people with wildlife backgrounds, and prospects are good. The Endangered Species Act, especially, makes it necessary for private industries such as logging and pipeline companies to hire their own consulting firms for information about potential operation sites. The information gathered can aid them in mitigating their proposed operations to stay within legal guidelines or in abandoning proposed projects before substantial sums of money are invested. Nonprofit groups also hire wildlife biologists for the same type of work, although their primary purpose may be somewhat at odds with that of private firms.

qualifications: Most employers prefer candidates to have a master's degree in biology or a related science, but they'll accept other candidates with applicable work experience.

characteristics: Someone with a great deal of patience and endurance will find this type of work right up his or her alley. Often it's necessary to hike across rough terrain at a killer pace, then wait for an animal that may never show up. You have to be physically fit and alert. You should enjoy camping, because you may be doing a lot of it, especially if you're working for a nonprofit organization.

Eric Meyer *is a wildlife biologist with a private consulting firm in New Mexico.*

How did you get the job?

Eric Meyer has a degree in wildlife biology, specializing in birds of prey. In the years after graduation from college he worked with a nonprofit group doing counts of migrating birds of prey. He found that job advertised in an environmental opportunities newsletter. His current job presented itself through contacts there.

What do you do all day?

Eric typically spends a tremendous amount of time in the field. For his current assignment he's checking on wildlife reproductive success and bird-of-prey density in the vicinity of a surface coal mine. Every project begins with research on the species. Then he packs his back-

> **DON'T GET DISCOURAGED IF YOU CAN'T FIND A FIELD OR RESEARCH PROJECT THAT INVOLVES YOUR SPECIALTY. BE WILLING TO DO FIELDWORK IN OTHER GENERAL OR SIMILAR SUBJECTS. GETTING THE EXPERIENCE IS WHAT COUNTS, AND BASIC SCIENCE SKILLS ARE OFTEN TRANSFERABLE TO OTHER FIELDS OF STUDY.**

pack with, among other things, his observational tools—items such as binoculars, notepads, checklists, and guidebooks— and heads out into the field. He may be out just for a day hike or may camp out under the stars for a few nights. He says he usually works alone or with one other person. The hours vary; projects such as a songbird sur-

vey can take from sunup until 9 or 10 at night. Does the job ever get boring? "Sometimes, especially in the fall and winter when you're in the cold and not many birds are passing," Eric says. "However, during the peak of the season the time passes quickly as you count some 600 birds in a day." Eric says that to do the job well, your identification skills must be sharp, and you get that only through experience. "You can't just remember it from a field guide.

Anybody can look up in the sky, but you need to know what you're looking at."

Where do you see this job leading you?

If he were to stay with a private firm, Eric says, the career path leads to wildlife crew leader at a larger company, and possibly to project manager, although that job would require a master's degree. Eric intends to go back to a nonprofit group or even to create his own. Ideally, he'd like to run a research station studying marine mammals in the Patagonia region of South America. He'd also like to write some articles for environmental publications.

12. District Wildlife Manager

description: In this job you're acting as a representative of the animals in the wild. A district wildlife manager is a type of wilderness cop who keeps track of designated species in his or her district and tries to avoid or ease the problems that can occur when people come into contact with them. The job is about protecting the animals and also about teaching humans.

In many instances, the encroachment of civilization throws ecosystems out of balance and some animal populations may need to be forcibly reduced or moved.

salary: Salaries start at about $24,000 a year.

prospects: The number of positions is pretty much fixed by geography and doesn't change very often. Colorado, for instance, employs 120 district wildlife managers and always has a waiting list of applicants.

qualifications: A bachelor's degree is usually mandatory, although the broad scope of duties leaves the field open to people of varied educational backgrounds. You need to have an understanding of environmental issues and of the underlying science of wildlife management, as well as a familiarity with the outdoors. Law enforcement training is required, including proficiency in the use of firearms. You should be physically fit.

characteristics: Since this is partially a law enforcement job, you must be able to deal with people and defuse confrontational situations. Wildlife managers often have to set their own schedules and priorities, so they should be self-motivated and organized. Although more than half the job is indoors—dealing with paperwork and instructional programs—there is also physically demanding outdoor work.

Patt Dorsey *is a district wildlife manager with the Colorado Division of Wildlife.*

How did you get the job?

It took Patt Dorsey eight years to land this job. She had a degree in wildlife biology, but was required to take additional written and oral tests to get hired as a wildlife manager. In the years just after graduation, she says, she spent a lot of time holding two jobs: one to pay the bills, and one as an environmental educator at a local park so she could build up her credentials and contacts.

What do you do all day?

On what might be considered a typical day, Patt trapped a bighorn sheep in the Rockies in the morning and was talking to a classroom full of students about the area wildlife that afternoon. The duties of a wildlife manager are, in fact, so varied that it's tough to pin

> **GET IN TOUCH WITH YOUR LOCAL DIVISION OF WILDLIFE AND ASK IF THEY HAVE ANY VOLUNTEER PROJECTS.**

down a typical routine. Patt works out of her home. She is on call 24 hours a day and deals with everything from the tracking and management of elk herds, to writing tickets for people fishing without a license. "The truth is," she tells us, "I deal with people more than wildlife. It's a hands-off approach, unless you're dealing with a problem." But there are problems: for instance, the elk herd annually needs to be culled. She tracks the ratios of bulls to cows and cows to calves, and that information is used to determine what types of hunting licenses are issued the following year. If a bear wanders into a populated area, she may have to tranquilize it and take it

back up into the woods. There's a golden eagle nest she's monitoring, but to see it she has to climb up the far side of a canyon. She's enlisted volunteers to help with that. On the day we called, she was preparing to address a county commission meeting about the impact of a pending land-use decision on the area's wildlife.

Where do you see this job leading you?

For now, she likes where she is, but Patt says she has considered going back to school and perhaps teaching adults to help people who otherwise might not have the chance to learn about wildlife.

13. Fisheries Biologist

(FEDERAL GOVERNMENT—U.S. FOREST SERVICE)

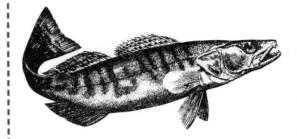

description: A fisheries biologist helps government agencies and private companies gather and evaluate the information that tells us if aquatic resources are being overtaxed or damaged. Broadly defined, the job involves dealing with anything that lives in streams, rivers, or lakes. Federal and state legislation, such as the Endangered Species Act, guides you as to what to look for, but in some government jobs you may have wide latitude in how to proceed.

salary: Entry-level salaries with a bachelor's degree are $19,000 to $20,000. Entry-level salaries with a master's degree will likely begin at $24,000 to $25,000. Salaries increase with experience and level of responsibility.

prospects: Environmental agencies report understaffing problems. Caseloads are high, but budgets are tight. An antiregulatory political atmosphere is not likely to improve the job market in this area, but better economic conditions could increase the number of projects whose impacts on the environment will need to be assessed. This may open entry-level trainee positions to help with the workload.

qualifications: Fisheries biologists are always trained scientists with four- or six-year degrees. Internships, volunteer work, and undergraduate field study will familiarize you with the work environment and help you make the contacts you'll need later.

characteristics: You'll need to be a self-starter. Because of the shortage of biologists, there isn't a great deal of leadership available at some agencies. Fisheries biologists are often innovative, spotting what needs to be done and implementing it themselves. You have to like the outdoors but not be afraid of paperwork. Expect to get wet.

Sheryl Bryan *is a fisheries biologist for the U.S. Forest Service, Department of Agriculture.*

- -

How did you get the job?

Sheryl Bryan made contact with the Forest Service while earning her master's degree. She enrolled in a cooperative education program in which she worked for four months helping a fisheries biologist and then went back to school to finish her degree. When she graduated, she knew she wanted to get a job in fisheries biology.

What do you do all day?

In the summer and fall, Sheryl says, she begins work at first light. Her day starts with a truck ride to a project site where she will supervise a crew collecting information about fish population, water chemistry, stream channel measurements, or even mayfly larvae—which live in the streambeds. In July through October, she works until sundown, as the days begin to shorten. In winter and early spring, Sheryl says, she is indoors, writing reports and evaluating the information gathered during the field season. These reports generally revolve around the impact of various land uses on the aquatic resource, and they are frequently shared with other government agencies with responsibility for issuing permits and enforcing environmental laws. Recently,

> **LOOK FOR ALTERNATIVE TITLES FOR YOUR JOB. AN EMPLOYER OR RESEARCH PROJECT DIRECTOR MAY CALL YOUR WORK BY A DIFFERENT NAME. FOR EXAMPLE, FISHERIES BIOLOGIST MAY ALSO BE LISTED UNDER WILDLIFE BIOLOGIST, FISH AND WILDLIFE TECHNICIAN, OR FIELD BIOLOGIST.**

Sheryl said she was finishing a proposal for a land swap with a private property owner. In this case, the Forest Service was interested in protecting a watershed area from development.

Where do you see this job leading you?

Sheryl believes that by showing initiative, she can carve out a career in a government agency, perhaps in administration. She admits, though, that her personal interests lead her toward teaching, which she could do through the Forest Service or in a school.

14. Field Biologist— "Plankton Police"

description: Where industry or hazardous material transport operations come into contact with coastal ecosystems, someone has to monitor the effect on the local food chain. The microscopic bottom of the food chain is plankton: the tiny plants and animals that support all living things in the oceans and coastal estuaries. A field biologist in this line of work monitors plankton levels for signs that oil spills or hazardous waste emissions might be affecting marine life.

salary: Salaries generally start at $20,000 to $25,000 a year.

prospects: In many areas the activity is mandated by federal law. Both the federal government and many states hire people for these positions. Job availability may vary with the level of industrial activity in a particular area. For instance, an increase in offshore oil production will require heavier staffing levels; a cutback may involve staff reductions.

qualifications: A marine biology degree is required, along with a working knowledge of chemistry, hydrology, and related fields. You should be comfortable on the water and be able to work with other scientists and technicians who collect overlapping data.

characteristics: This is really a team effort. There are lots of samples to be collected, lots of lab work to perform, lots of "housekeeping" chores to be shared. The essential element here seems to be the ability to work in tandem. Monitoring something as small and pervasive as plankton is a massive job.

Jason Duet *is a senior field biologist with the Louisiana Department of Wildlife and Fisheries in Baton Rouge.*

- -

How did you get the job?

Jason Duet got the bug for this type of work during his college summers. For three years running, he landed a biological technician's job with the National Marine Fisheries Service. Through word of mouth he heard that the big deep-water oil port in Louisiana, LOOP (Louisiana Offshore Oil Port), was hiring in its environmental lab. As it turned out, the position was a state job. He had the right schooling and experience, so he got the job.

> **JOB INTERVIEW TIP: LIST YOUR STRENGTHS. TELL THE INTERVIEWER EXACTLY WHAT YOU CAN DO FOR THE COMPANY. TELL HIM OR HER OF A UNIQUE SKILL, EXPERIENCE, OR QUALIFICATION YOU HAVE THAT WILL BENEFIT THE COMPANY OR AGENCY.**

What do you do all day?

There's lots of lab and computer work: compiling the data and entering it into a database. Jason says he goes out on seven or eight field trips per month, sometimes offshore on an 85-foot converted trawler, some-times into marshes in smaller boats. The field days can go as long as 16 hours, but he loves being on the water. The members of the "Plankton Police" collect plankton and nekton samples. They also measure oxygen, nitrogen, chlorophyll, phosphorus, and potassium levels at sea and along the inshore oil pipeline. But there are less glamorous days too: Jason recently disposed of five- and six-year-old plankton samples. That meant straining old plankton out of large bottles of formaldehyde. While everyone else was offshore on a field trip, he was left working with loads of 16-ounce jars in 100-degree heat, wearing a respirator and goggles. Field trips can be rough, too. Jason remembers one offshore outing last year when the weather was so nasty that most of the 13 team members were seasick.

Where do you see this job leading you?

Jason thinks he'd like to work next for the federal government. He says pay and promotion paths are a little better. He could head to private business, but he likes this end of it. "I'm a glutton for punishment," he says.

15. Fish Hatchery Worker

description: Fish hatcheries, or "fish farms," as they're sometimes called, produce many types of fish, from trout and catfish to shrimp and oysters. They supply stocking programs, restaurants, and markets. A fish hatchery worker's responsibilities include the rearing and feeding of fish, and the cleaning of troughs, ponds, pens, and other fish or wildlife enclosures. It's a hands-on, dirty job as these workers collect data, samples, and specimens. They operate pumps, fish loaders, and other types of mechanical equipment. Often it's also the fish hatchery worker's job to log data and keep records. These workers also remove barriers from streams and maintain fish screens, ladders, and traps. At certain times of the year they take eggs from the fish, or culture the eggs in smaller ponds.

salary: Hatchery workers start at about $20,000 a year, and at the senior level may top out at around $55,000.

prospects: It is a very competitive field, with sometimes 50 to 60 people vying for the same job.

qualifications: State-operated hatcheries usually require a minimum of one year of college with at least some courses in the biological sciences. You should have basic knowledge and experience in fish and wildlife conservation. It's recommended you get schooling in fish health, reproduction, breeding, and genetics. You should also be familiar with harvesting techniques and water quality management. A background in business management may give you the edge to quickly move up the ladder to hatchery manager. State hatchery jobs usually require taking a written test.

characteristics: A fish hatchery worker needs to be in good physical condition and have the willingness to work unusual hours, weekends, and holidays. Keen observation, tact, reliability, and the willingness to do heavy manual labor are also prerequisites for the job.

Bruce Barngrover *is a fish hatchery supervisor.*

How did you get the job?

Bruce Barngrover didn't get started in this field until he was 36 years old. He'd been a salesman and was looking for a career change. In his search he heard about hatchery jobs. He'd always been interested in outdoor activities, and on vacation trips he often stopped at hatcheries. Bruce says once he decided this was what he wanted to do, he sought out what he needed to study and talked to the right people. He then took the required state exam, and was hired soon after.

What do you do all day?

Bruce is now a supervisor in charge of four state-operated hatcheries that raise mostly salmon and trout used to stock state ponds, rivers, and streams. Much of the work is hard manual labor. "There is a lot of maintenance work to do in the facility and grounds. There's

> **GET EXPERIENCE DOING SEASONAL STATE WILDLIFE CONSERVATION WORK. CHOOSE A SCHOOL WHOSE PROGRAMS OFFER FIELD STUDY WITH RESEARCH PROJECTS OR AGENCIES.**

feeding of the fish and cleaning of their ponds," says Bruce. The staff also gather and collect the fish and transport them to other regions. They closely monitor water quality, fish production, population counts, and overall health of the stock. In certain seasons, fish hatchery workers are busy taking eggs from the fish to hatch them in incubating facilities. Perhaps the most important responsibility of the hatchery worker is keen observation of the stock. Disease in a pond can wipe out millions of fish. When a hatchery worker or fish culturist suspects that there's something wrong, he or she calls in a fish pathologist or biologist to make an analysis and recommend treatment.

As a supervisor, Bruce spends a lot of time in the office doing administrative work, but he's always ready to tackle an emergency. Recently, Bruce was fielding questions from the media regarding a large break in a nearby dam. One hatchery sits under the dam. A spillway had broken, releasing tons of water and causing a turbulent situation. Luckily, the emergency was contained and loss of fish was minimal.

Where do you see this job leading you?

As a senior administrator, Bruce is at a level where he's ready to retire. Asked what makes the job interesting, he says, "You're working with a live crop, you're working with fish. What's so interesting about the job? They're alive."

16. Nature Preserve Manager

description: A nature preserve manager is responsible for the overall maintenance, protection, and restoration of ecologically sensitive or threatened sites. Nature preserves may be owned by organizations such as the Nature Conservancy, one of the largest and most well-known nonprofit groups dedicated to the preservation of natural areas throughout the world. Supported through donations, the Nature Conservancy identifies and acquires endangered land parcels and maintains them as natural preserves. Preserves are also sometimes publicly owned lands maintained by local governments—such as a city, county, or state—or by federal agencies such as the Bureau of Land Management. The preserve manager is the overall caretaker of the sanctuary. His or her job includes the protection of the natural habitat and wildlife and the management of personnel and volunteers. Responsibilities include determining the biological objectives of the preserve and the necessary conservation or restoration methods to achieve them.

The job may involve deciding how to control exotic vegetation or developing a long-range wetlands restoration project. Preserve managers and their families usually live in or near the sanctuaries, which are often located in somewhat remote areas.

salary: Salaries depend on the size and location of the preserve as well as on the complexity of the job at that particular sanctuary. Salaries can range from $23,000 to $48,000 a year.

prospects: These jobs are difficult to find. There is low turnover and these are not usually entry-level jobs. Openings are usually filled with experienced conservation or wildlife managers.

qualifications: To work in a nature preserve, one should have degrees or experience in conservation or wildlife management or one of the biological sciences. Business courses are also helpful.

characteristics: Preserve managers need to be extremely responsible and self-motivated. They are problem solvers who must formulate and offer creative and economical solutions. They also have to possess the type of personality that would enable them to build a communicative rapport with surrounding landowners.

Jim Bergan *is a nature preserve manager/supervisor in coastal Texas.*

How did you get the job?

Jim Bergan has a bachelor of science degree in forestry/wildlife management and got his master's and doctorate degrees from Texas Tech University, where he studied waterfowl populations in South Carolina and the Playa Lakes regions of Texas and New Mexico. While in college Jim worked as a co-op student for the Indiana Division of Wildlife. He also worked at the Savannah River Ecology Laboratory in South Carolina. After graduation Jim took a job with the state of Florida as a waterfowl program supervisor, but for personal reasons he decided he wanted to move back to Texas. He says he just "put out some feelers" and heard about a job with the Nature Conservancy. Hired initially as a manager of a single preserve, he presently is the Coastal Texas Land Steward in charge of several protected areas.

What do you do all day?

Jim's job is a big one—he's in charge of more than 7,000 acres of one nature preserve, along with three other smaller preserves. The marsh complex is one of the most important areas in Texas in terms of waterfowl, shorebirds, and other species. Jim was hired when the Nature Conservancy first acquired the property.

His first job was to turn a dilapidated hunting lodge overrun by animals into office space and lodging for staff and research crews. Not only did the hunting lodge need work—so did the entire property. It took a lot of sweat, muscle, and grit on the part of Jim, his staff, and a corps of volunteers to clear areas, build roads, and post boundary signs. Jim and his family now live in a house on the edge of the reserve where alligators take shade under the carport and water moccasins are occasionally found in the front yard.

One of the first challenges Jim faced was trying to restore the wetlands to their original state. He and his staff are also experimenting with growing rice, which provides feeding and roosting habitat for wintering birds and other wildlife. On any given day, Jim might be either in the office, checking a research team conducting bird surveys, or meeting with an adjacent landowner. He also takes visitors on tours of the preserve. He says that he now spends 85 percent of his time in the office writing proposals and reports, revising guidelines, and talking on the telephone.

Where do you see this job leading you?

Jim has moved up the ladder rather quickly as the area in his charge has grown from 3,000 to more than 7,000 acres. He hopes to continue moving up the career ladder with the Nature Conservancy.

> NATURE PRESERVES AND PARKS ARE ALWAYS LOOKING FOR VOLUNTEERS. VOLUNTEERS DO EVERYTHING FROM HELPING IN CLEANUP PROJECTS TO CREATING NEW NATURE PATHS WITH WOOD CHIPS.

17. Air Quality Inspector
(STATE GOVERNMENT)

description: An air quality inspector ensures that man-made emissions into the atmosphere do not violate provisions of state or federal air pollution laws, including the Clean Air Act of 1990. This isn't just a matter of testing the air; the inspector is a government regulator who writes air emission permits and reviews permit applications, most of which are from businesses or factories that discharge particulate matter or chemicals into the air. Inspectors are also required to investigate citizen complaints and alert the appropriate law enforcement agencies if violations are discovered. They make on-site visits to factories, businesses, and sometimes even private homes to examine equipment, records, and industrial chemicals to make sure operations are in compliance with pollution laws. As an inspector you may have to supervise teams of technicians who will collect the information you'll need.

salary: Entry-level salary for air quality inspectors with engineering degrees is $33,000 per year.

prospects: This area underwent a growth spurt with the signing of the Clean Air Act. Some offices doubled their staffs in the first years of the 1990s. Some inspectors, especially those with engineering degrees, move on to private consulting firms. Colleges and universities are responding with engineering degree programs that specialize in the environment.

qualifications: Chemical or environmental engineering or environmental education degrees are preferred.

characteristics: You will need a working knowledge of chemistry and various applications of chemicals in industry, as well as a firm background in how these industrial chemicals behave in the environment.

Mike Parkin *is an environmental engineer specializing in air quality control.*

How did you get the job?

Mike Parkin always thought he'd be working with water, not air. He was wrapping up his master's degree in civil engineering with a course emphasis on water chemistry and environmental concerns when he spotted his current job posted on a bulletin board at the university placement office. He says his resume and a little luck did the rest. When he landed his job, he says, many schools didn't offer specific air pollution programs. He had to adapt what he'd learned about water to his new position. Making that adaptation was a kind of on-the-job training.

What do you do all day?

It's a mix of office work and on-site inspections. In the winter there's more paperwork, in the summer more inspections. A typical factory inspection will take up to two days, depending on the size of the facility. The paperwork review will necessitate a full day in the office. Citizen complaints sometimes require more field time per case. Mike says his office also handles open-burning complaints, which sometimes involve going to private homes to tell residents they can't burn trash in their backyard. People have been known to get angry, which Mike says can be a difficult situation for an unarmed environmental engineer. He used to go out on calls alone, but since the

> **REMEMBER THAT BODY LANGUAGE COUNTS.**
> **WALK INTO AN INTERVIEW LOOKING CONFIDENT, WITH YOUR HEAD UP, SHOULDERS SQUARED, AND READY WITH A FIRM HANDSHAKE.**

office now has more employees, he's usually accompanied by two or three technical engineers.

Because most of his work involves inspecting factories in a 17-county area, Mike says he's learned a lot about industry. "Every possible thing you can think of, I've seen it made—boat motors, batteries, and baby diapers. It's an interesting job."

Where do you see this job leading you?

This work is excellent preparation for environmental consulting. People who've worked for the government in this field are especially valuable to private industry because they know how to handle the permit application process. Mike, however, is not necessarily interested in that. Right now, he says, he gets satisfaction seeing people and businesses switch from using toxic compounds to trying less harmful substances and seeing them put pollution control equipment in place.

18. Air Quality Engineer

(PRIVATE CONSULTING FIRM)

description: Dealing with air pollution is more than a matter of taking air samples. An air quality engineer is the person who looks at all the information available at a potential pollution site to figure out what is happening and how to avoid or mitigate the problem. In many ways he or she is a detective, looking for clues that will point to the truth, even when there's no "smoking gun" at hand.

salary: Salaries start at around $30,000.

prospects: This is a hot field. Amendments to the 1990 Clean Air Act have resulted in binder-size permit and documentation packets. Help is needed on both the regulator and industry sides of the issue. Some firms have doubled their air quality staffs in recent years.

qualifications: A chemical engineering, civil engineering, or related degree is needed for this position.

characteristics: An air quality engineer has to be intuitive but methodical. In this job you're identifying areas that need to be tested and examining things, such as supply records, to extrapolate how much pollution might be generated in any given area. When records aren't available, you need to interview employees and suppliers. Often there are several projects in progress at the same time; flexibility is a must.

John Kingsley *is an air quality engineer with a private consulting firm.*

How did you get the job?

John Kingsley targeted consulting firms as he approached graduation, got an interview, and made the cut. Had the company turned him down, he says, he might have gone to work for the government, possibly with the Environmental Protection Agency's hazardous materials response team.

What do you do all day?

John tours everything from naval bases to power plants, iron mines, sneaker factories, and paper mills. His inspections are walking tours, usually carried out in the company of the plant's environmental manager. John tries to pinpoint all the possible sources of pollution, then contracts for samples to be taken. He combines the information from the samples with data from files and interviews with plant employees to get a picture of what's happening. If necessary, he guides the plant management through the lengthy process of filling out state permit applications.

IN A JOB INTERVIEW, BE READY TO ASK YOUR OWN QUESTIONS, SUCH AS "WHAT WOULD A TYPICAL DAY BE LIKE?" OR "WHAT IS THE OVERALL PHILOSOPHY OF THE COMPANY?" YOUR QUESTIONS WILL SHOW THE INTERVIEWER YOUR SINCERE INTEREST IN THE JOB.

Where do you see this job leading you?

John is now working toward a master's degree in environmental engineering in order to secure promotions in this line. He'd like to be a project manager, which would give him the opportunity to deal with all aspects of the company's relationship with a given client. To limit the amount of traveling he does, he'd eventually like to work as an environmental manager in a single plant.

19. Indoor Air Quality Technical Specialist

description: Taking air samples, counting light bulbs, and measuring vents are among the varied tasks an indoor air quality technical specialist will perform in an average day in order to determine the air quality of a building. These workers must be motivated and willing to take on any chore, whether it is climbing up on a rooftop or crawling through a small ventilation space. Because the amount of information collected at a site is so massive, air quality technical specialists usually spend their days working in the field with three or four other experts.

But gathering the data is only half the job; the information must then be compiled into a database where it can be interpreted. Therefore, technical specialists spend long hours in front of a computer. Although most find this task the most tedious part of the job, it is considered equal in importance to their fieldwork.

salary: Someone starting out in this industry with a bachelor's degree can expect to draw $25,000 or more a year. In order to gain a position with greater responsibility and more pay, such as an industrial hygienist, some type of graduate work must be completed. A person with a master's degree could expect to make about $30,000 a year. Someone with 10 years of experience could expect to make at least $50,000 per year.

prospects: The prospects are good in the up-and-coming field of industrial hygiene. More and more people are becoming aware of the effects that their working environment has on them. This raised awareness can only lead to more job openings in the field.

qualifications: A bachelor's degree in one of the sciences, preferably environmental science, is required for this type of work. However, any degree in the sciences, coupled with an internship in the field, is an acceptable starting point.

characteristics: A person working as an indoor air quality technical specialist generally has a high energy level and a genuine concern for the consequences that the activities of society have on the environment. Because fieldwork is usually embarked upon by a team of people, technical specialists are often surrounded by others with varied experience levels and areas of expertise. These workers generally enjoy being in this type of setting and are able to work well with others.

Dana Hanson *is an indoor air quality technical specialist with an environmental health and consulting firm.*

How did you get the job?

When Dana Hanson headed off to college, she didn't plan on seeking a career in environmental work. But after taking an elective in environmental studies, she decided that the environment offered her an interesting career path. Dana says that now that she has begun working in the field, she is glad that she chose to pursue this path. During her junior year, while she was working toward her bachelor's degree in environmental studies with an emphasis on policy, she set out to find an internship in the field. She was interested in working in recycling, but when she called the department of environmental management, they didn't have any internships available. One of her professors suggested that she call the local health department.

After her internship with the health department, Dana was hired to work with a lead poisoning prevention project. When the funding for the project she was working on ended,

ALWAYS FOLLOW UP JOB INTERVIEWS WITH A WRITTEN THANK-YOU NOTE. THE NOTE REMINDS THE INTERVIEWER ABOUT YOU AND IS A GOOD OPPORTUNITY TO UNDERSCORE YOUR SUITABILITY FOR THE JOB. IT'S ALSO A CHANCE TO INCLUDE SOMETHING ABOUT YOUR EXPERIENCE THAT YOU MIGHT HAVE FORGOTTEN TO MENTION IN THE INTERVIEW.

one of her co-workers suggested that she try to find work with a consulting firm called Environmental Health and Engineering. "I was really lucky. I sent a letter and resume and they just happened to be hiring."

What do you do all day?

Dana's first day at Environmental Health and Engineering gave her a good idea of the variety of tasks she would be expected to perform. "My first day on the job was in a snowstorm, and my supervisor was trying to figure out how well the instrumentation would work in really cold weather, so he had me crawl up on the roof and we had to test 'data loggers' and it was my first day. I'd never seen a data logger and he sent me up on the roof. It's basically been that sort of stuff from day one." When Dana goes out with a team into the field to conduct a study, she knows she will spend most of the day on her feet because there is so much information to be gathered and so little time. She usually starts by talking to the building manager to find out what kind of activities take place in the building. The rest of her time will more than likely be spent running through hallways counting the number of people occupying the building, or taking a measurement of the humidity in the building.

After a week in the field there is always a need to com-

pile the information. This can be a tedious process, but it is what makes all the time spent out in the field pay off. While sorting through the data is not Dana's favorite part of the job, she knows that it is something that has to be done in order to meet her overall goal, which is to make a difference. "Sometimes you get bogged down in administrative details and stuff and you sort of scratch your head and say, 'Is this making a difference?'—but I think in the end you're sort of working toward that goal and you realize that it all will sort of fall into place."

Where do you see this job leading you?

While Dana enjoys the work she is doing now, in order to advance, she feels the need to further her education. Dana plans on pursuing her master's degree in industrial hygiene. From there, she says, she will either continue working for a consulting firm, look into government work such as writing policy, or find a job creating health and safety programs for corporations.

20. Drinking-Water Quality Control Director

(MUNICIPAL WATER SYSTEM)

description: Just as the job title implies, the water quality control director helps ensure that drinking water stays drinkable. This involves constant testing, sifting through data, and ordering measures to correct potential problems before people get ill. In this job, you are expected to know the science cold, to be able to identify problems before they become serious and, perhaps most importantly, to deal with the public's concerns and worries about the safety of what comes out of their taps at home. The job is found in government at the municipal, state, and federal levels, as well as in privately owned water companies. Sometimes local health departments perform the function as monitors of a water system.

salary: Salaries range from $22,000 to $50,000, depending on the size of the water system.

prospects: Concern about bacteria and waterborne pathogens makes this a hot area. Opportunities exist at all levels of the water monitoring process. Organizations managing water systems are under federal mandate to prove they're safeguarding the water supply.

qualifications: A bachelor's degree in the appropriate sciences will get you in the door, but you'll need a master's to advance to the head of any sizable department. Writing and computer skills are also required.

characteristics: Because you'll be the point person for the public's complaints and questions about water quality, verbal and written communication skills are critical. You'll have to be able to either interpret raw data or construct the computer models to help you do so. Administrative skills are also part of the game.

Doreen Bader *is the drinking-water quality control director for the city of New York.*

How did you get the job?

Doreen Bader graduated from Southampton College with a degree in marine biology and nowhere to go. There just weren't many jobs in marine biology. She happened to attend a Women's Club meeting where a drinking-water quality supervisor was giving a speech. Since Doreen always carried her resume with her, she handed him one and followed up with a phone call. Coincidentally, a job had just opened up. She was called in for an interview and hired one day before the implementation of a city hiring freeze.

What do you do all day?

"Help! The water coming out of my faucet is brown! Is it safe to drink?" "Is there lead in my drinking water? How can I get it tested?" Doreen spends about 40 percent of her day on the phone answering questions such as these from the general public—customers worried about the color or taste of the water, or concerned about a problem they heard about in the media. She makes sure that the drinking water for eight million New Yorkers is safe.

The first thing Doreen does when she gets into the office is check her computer for the daily test data. In a crisis, she may have to go into the field and supervise the flushing of a section of the distribution sys-

> ASK TO TAKE A TOUR OF TREATMENT FACILITIES OR WATER RESOURCES PROJECTS, AND GET TO KNOW AS MUCH AS YOU CAN ABOUT THE DRINKING-WATER SYSTEM IN YOUR COMMUNITY. DON'T FORGET TO BRING ALONG YOUR RESUME.

tem, or perhaps order a chlorine "shock" to one section of pipeline. Summer can be the busiest time; that's when water is warm, providing perfect conditions for bacteria to thrive, making this a time when a drinking-water quality person must be most diligent in moni-

toring test data. It's also the time of year when water alerts are usually issued. If the public is under a "boil water" alert, Doreen has to work seven days a week until the problem is solved. She's a management person, and that means lots of staff meetings.

Where do you see this job leading you?

Doreen likes working for the city. She'd ultimately like to transfer to one of the upstate reservoirs, just to get away from urban living. There are opportunities in private consulting firms, but she'd be more inclined to go back to school for her Ph.D., which would increase her chances of doing more microbiology research.

21. City Director of Water Conservation

description: As any city's director of water conservation, some of your time may be spent out of the office crawling through buildings to inspect plumbing systems. But it's likely that more of your time will be spent in the office coming up with new programs to provide the public with incentives to conserve water. Once these incentives are in place, you will concern yourself with their administration to ensure that they work as planned. This can be done by sorting through mail, answering calls from both angry and pleased members of the public, writing proposals, attending meetings, and even speaking at various engagements. But whether you are working indoors or out, your focus—conserving water—will stay the same. You will find reward in the effect you're having on the public's general welfare, as well as in the role you are playing in preparing for the future.

salary: Because the position of a director is considered to be lower- to middle-management, the pay is often high. However, the range of salaries can be about as broad as the range in cities. On the low end, you will bring in about $45,000 a year. Larger cities may pay somewhere in the range of $65,000.

prospects: There is good news for those wanting to enter this field: the water-efficiency field is blossoming in many parts of the country in the same way the energy field did some 10 to 15 years ago. However, there is probably only one spot in every city—if that—for this particular position. Timing is everything.

qualifications: Although this is a city government position, it is not necessary to have previous experience working in government to land this job. However, it is imperative that someone coming in as the director have past experience in some sort of energy or water management. A bachelor's degree in environmental science is an ideal beginning for this type of career, and working in a water or energy consultation firm is a logical second step.

characteristics: For work as a director of water conservation, you must be motivated, well organized, and have a talent for problem solving. The problems you deal with daily, both managerial and technical, must be solved quickly and decisively. A fear of public speaking will not sit well in this position; there will be numerous occasions where you will be called upon to speak to the public in order to raise awareness or to speak in front of the city council about a proposal. The ability to write well is also a must, given all the proposals that you will generate.

Warren Liebold *is director of water conservation in New York City.*

How did you get the job?

After Warren Liebold set out looking for a job, he found that his bachelor's degree in biology, along with a large amount of work toward a master's in marine biology, wasn't much help. Consequently, Warren ended up working various jobs—as a teacher, a food co-op manager, and a Sierra Club staffer—before he found a job with an engineering firm.

The position as a junior engineer allowed Warren to work on an energy conservation program that the company had been contracted to oversee by the city of New York. As a junior engineer, Warren became familiar with the mechanical end of energy use by reviewing energy-efficiency analyses of different New York City buildings. "I had to start understanding air-conditioning systems and heating systems and windows and roofs, and understand how you calculate energy savings by making those things more efficient. I also spent lots of time crawling through steam tunnels and hoping that I was the only warm-blooded thing there at the time," he says.

Along with the new job, Warren decided to go back to school, where he began graduate work toward a master's in energy management. This background, coupled with volunteer work, prepared him for the challenges he would soon face as New York City's Department of Environmental Protection director of conservation. "I had been very active in energy-efficiency programs through my volunteer activities and had a technical background through my job. I was more than active enough in the energy-efficiency field, and I was working on public service commission cases, so I had a reasonably high profile." Not only had Warren gained a high profile through his volunteer work, but he had also developed a working relationship with the city's Department of Environmental Protection commissioner, who brought him on board as the director of conservation.

What do you do all day?

Warren spends a large amount of his time in the office with administrative duties: writing and editing proposals, working on speeches, and sorting through mail and phone messages. Each week, at least three of these in-office days will be interrupted by an out-of-office meeting or speaking engagement. Much of the time in the office is spent in the administration of conservation programs. Warren averages about two days out of every month doing fieldwork such as going to the future site of a housing project, where he might check the plumbing systems and identify possible conservation problems.

"It's never boring," he says. "It's never exactly the same. It's not a matter of me going and doing the same thing every day."

But it's more than the exciting pace that keeps Warren interested in his job. "You really can feel that you are making a difference. You are involved in getting something positive done in society, and I think that is also very worthwhile."

Where do you see this job leading you?

Warren says he sees his career leading him to continue doing the same type of work, but possibly in a different setting. Working for a water utility in a different city or for a water conservation consulting firm might be the next step for him.

> **THE BEST WAY TO GET EXPERIENCE IN THIS FIELD IS TO GET AN INTERNSHIP WITH AN ORGANIZATION THAT DEALS WITH ENERGY OR WATER CONSERVATION. THE PHONE BOOK'S YELLOW PAGES, UNDER ENGINEERING OR CONSULTING FIRMS, IS A GOOD PLACE TO START WHEN LOOKING FOR THESE SORTS OF JOBS.**

22. Aquatic Environmental Scientist

description: Researching water quality is the main focus for many aquatic environmental scientists, but there are numerous ways this can be done. Some go about this chore by collecting and running experiments on different life forms in the water, such as algae; others look toward the nutrient movements of a body of water; and still others study the water by looking at a computer-generated model. These computer casts are created to closely mirror a body of water and its specific conditions, such as the temperature of the air above and around it, the movement of algae, and different levels of chemicals within the water. But whether aquatic scientists are working near the water or at a computer, they are always striving to use their own findings and the findings of others in the field to learn more about bettering water quality.

salary: Someone starting out in this field can expect to make $30,000 to $40,000 in the first year. After two years of experience you will be eligible to take on more responsibility, which could raise your salary to between $35,000 and $50,000 a year. A senior environmental scientist can make up to $64,000 and a supervising professional can bring in up to $70,000.

prospects: Currently, the chances of finding employment as an aquatic environmental scientist within a government agency are slim because of the trend to downsize. However, jobs in this field can be found in the private sector; for example, environmental consulting firms often hire aquatic scientists.

qualifications: An aquatic scientist will need to go beyond a bachelor's degree in order to get work in the field—either a master's or a doctorate in the sciences. A degree in engineering, ecology, or any type of environmental science is helpful to someone seeking work in this area.

characteristics: Especially if you are working on computer models, there will be long stretches of working by yourself; therefore, an aquatic scientist should be self-motivated. Aquatic scientists are generally also inquisitive people.

Tom James *is an aquatic environmental scientist.*

How did you get the job?

Tom James first developed his love for the outdoors while hiking and camping as a young boy in the Boy Scouts. In college he decided he would pursue that interest by preparing himself to work in the environmental field. He earned a double degree in zoology and psychology from Duke University and followed this with a master's in biology and a doctorate in ecology.

During the doctorate program, Tom was introduced to computer models and their role in the environmental field. After gaining his doctorate, he began work at the Stennis Space Center in Mississippi, where he produced his first model. The model showed the production of methane in the Everglades.

Through contacts he made while getting his doctorate, Tom secured his current position at the Okeechobee Systems

ONE WAY TO FIND OUT ABOUT WHAT IS GOING ON IN THIS FIELD AND HOW TO BREAK IN IS TO ATTEND A NATIONAL ORGANIZATION MEETING, SUCH AS THOSE OF THE NORTH AMERICAN LAKE MANAGEMENT SOCIETY. ANNOUNCEMENTS FOR THESE TYPES OF MEETINGS ARE USUALLY POSTED ON COLLEGE CAMPUSES. IF YOU ARE HAVING TROUBLE LOCATING ONE, TRY ASKING A PROFESSOR.

Research Division in south Florida. There he is studying a model of Lake Okeechobee and working to find a way to reduce the algae growth and the phosphorus concentration in the largest lake in the southeastern United States.

What do you do all day?

Because Tom's work centers around a computer model of the lake, he spends most of his time indoors sitting in front of the computer. "Sometimes I feel like I am a computer person, and you get lost in that. It's easy to get lost in playing with the computer. It may be work, but sometimes it's play."

Despite the number of hours Tom spends using a computer, he says he is still able to get a sense that he is working with the environment when he writes the results of his studies. "Getting back to the writing and trying to put your ideas on paper—that gives you more of a sense of contributing and being in the real world."

Tom is able to witness the far-reaching effects that his studies have on the industry when he gets requests for reprints from as far away as Austria, Italy, and South America.

"Probably one of the nicest things is when you have people contact you and request reprints, because you know that they have seen your work and are interested in it enough to ask you for a reprint." But, Tom says, the most gratification will come when he is able to see his work pay off on Lake Okeechobee. "Eventually we're hoping that this particular model will help us in providing information to our agency that will tell them how better to manage the lake. That's very gratifying as well, but we're still waiting to get to that."

Where do you see this job leading you?

In the near future, Tom says, he hopes to advance to a senior position at Lake Okeechobee. As far as long-term goals, he would eventually like to go back to the academic setting and work as a college professor.

23. Wastewater Engineer

description: A wastewater engineer helps design and supervise the construction of systems to control runoff and sewage. In this field you are attempting to minimize the impact of potentially damaging water pollutants. Your employer is likely to be a private consulting or construction firm whose clients could range from a small developer who's building a parking lot near wetlands, to a city or state government that is building or modifying a large sewage-treatment plant. This is a job that matches the technical skill of a civil engineer with the real-life problem of protecting water supplies and aquatic environments. A wastewater engineer works in close tandem with electrical, instrumental, and structural experts who may be unaware of or lack expertise in dealing with effluent.

salary: The salary for a wastewater engineer starts at about $30,000 a year, but success over the course of a long career can lead to a six-figure salary.

prospects: Two major factors affect this type of work: development and changing environmental regulations. For instance, in a rural area that suddenly begins to fill with people, sewage-treatment plants must be constructed from scratch, or an older system may have to be upgraded. In addition, when laws are changed that affect water quality protection in a local environment, construction projects may suddenly be faced with having to design or redesign storm-water runoff systems. This is a growing field; the jobs are out there for qualified candidates. Most sewage-treatment work involves expanding or changing the capacity or treatment process at existing plants.

qualifications: Applicants for a job in this field need an engineering degree and a strong background in environmental science.

characteristics: Wastewater engineers must be practical people, able to balance the mechanical demands of construction with the process demands of wastewater treatment and effluent control.

Christine deBarbadillo *is a wastewater project engineer for a private consulting firm.*

How did you get the job?

Christine deBarbadillo has a bachelor's degree in civil engineering and a master's in environmental science. A classmate got a job for a private consulting firm and told Christine they were hiring entry-level people, so she applied. That was seven years ago.

> **RESEARCH A POTENTIAL EMPLOYER. BEFORE AN INTERVIEW, ASK FOR ANY CORPORATE PUBLICATIONS, AND CHECK BUSINESS AND TECHNICAL JOURNALS.**

What do you do all day?

Christine follows a project from start to finish. She says it's the best thing about her job, but it also means that she has long stretches of indoor work, interspersed with extended periods of on-site supervision. She recently completed a two-year refitting of a sewage-treatment plant, a project that had her on site every day. This past season she designed a system to pump off foam from a sewage plant aeration tank. Recently, she was beginning a feasibility study to see if a local sewage-treatment plant could improve its effluent in order to apply for more stringent permits. She says she's been lucky in that most of her projects have been close to the branch office, which is near her home, but others in her field have had to make extended out-of-town trips. Not all jobs are equally interesting either, she says: people in her field could find themselves doing a grid-work for a parking lot storm sewer. The challenge of the jobs increases with your level of experience.

Where do you see this job leading you?

In the company she works for, Christine sees a clear career track from project engineer to project management to associate of the company or branch manager. In the long run, though, she'd like to have her own consulting business.

24. Wastewater-Treatment Plant Operator

description: As a wastewater-treatment plant operator, you oversee the running of pumping stations and the maintenance of the equipment that treats wastewater and removes and treats suspended solids in that water. This enables sewage to be released back into the environment without causing excessive damage. When things are running normally, it can be a quiet routine. In emergencies, you'll need the knowledge and expertise to get things back on track as quickly as possible.

salary: Starting salaries range from $17,000 at smaller facilities to $27,000 at larger ones.

prospects: In the area of water quality management, wastewater treatment is the largest field. Wastewater-treatment plants are everywhere, and technicians are needed both to operate the plants and to monitor and test wastewater content and treatment efficiency.

qualifications: A two-year college degree and state certification are required in many places. Knowledge of the underlying scientific principles of the process and of hydro-engineering is also a plus. Basic computer knowledge is essential.

characteristics: Day in and day out, it is a repetitive job. You must be able to pay attention to detail, as the job requires keeping up with test samplings and data readings. Being in charge of a plant also requires the traditional managerial skill of diplomacy and the ability to delegate responsibilities. It's also a team effort, so you've got to be a team player.

Bill Horst *is the supervisor at the municipal wastewater-treatment plant in Lancaster, Pennsylvania.*

- -

How did you get the job?

Bill Horst was appointed to his current job by the city administration. He's been a city employee in Lancaster since 1971. He started as an entry-level plant operator back in 1971 and moved up through the ranks as the plant expanded under the requirements of the federal Clean Water Act.

Bill is not a college graduate, but he took a community college course in operations and treatment of wastewater-treatment plants in preparation for his certification by the state. It has taken that kind of resourcefulness and initiative for him to move up the ladder quickly. Now he's in charge of all the town's wastewater operations, including pumping stations, solid-waste facilities, and treatment facilities.

What do you do all day?

Bill's life is that of a boss. "I'm constantly on my computer," he says, "looking at budget expenses." As a manager he oversees the plant's budget and a staff of 48 people who operate the treatment facility and the pumping stations. He signs off on operation papers that have to be sent to the EPA and watches for changes in the "biomass" (the biological mass of microorganisms used to treat the wastewater, as opposed to chemical treatment).

Water from this plant ultimately discharges into the Chesapeake Bay, so regulations are stringent and the system requires closes monitoring. Bill makes sure that samples are extracted and analyzed on a regular basis. Some of the testing is done on the premises, and other samples are sent to larger commercial laboratories.

Bill says that although the job doesn't change much from day to day, he has to be ready to handle any problem that comes along. To keep up to date, he attends several conferences a year.

Where do you see this job leading you?

Bill is in this for the long haul, but he says there have been people with Ph.D.s and engineers who have worked at the plant on their way to jobs in private industry.

> **EVEN THOUGH A COLLEGE DEGREE ISN'T REQUIRED TO WORK AS AN OPERATOR IN A WASTEWATER-TREATMENT PLANT, YOU SHOULD SHOW INITIATIVE BY TAKING AT LEAST A COMMUNITY COLLEGE COURSE IN ONE OF THE RELATED SCIENCES.**
> - - - - - - - - - -

25. Water Pollution Investigator

description: A water pollution investigator is a detective of sorts. The job involves collecting evidence that will prove the guilt or innocence of the alleged polluter who is under investigation or suspicion. This means collecting samples and testing them on an almost continuous basis. The investigation "targets" can be regular monitoring jobs, complaint follow-ups, or surveillance of chronic violators. The work can be repetitive and physical. In this job you may find yourself walking a streambank in a remote forest or climbing down a sewer opening in the middle of a city street. This work is done at all hours of the day in all sorts of weather.

salary: Salaries start in the high-$20,000 to mid-$30,000 range.

prospects: In this field, prospects are improving, especially for degree holders, as various public agencies that do this type of work upgrade their field forces.

qualifications: Most employers now require a science degree of some sort for this job. People trained in biology and/or chemistry seem to be in demand.

characteristics: In this job you have to be ready to go wherever the water is, and that isn't always pleasant; a lot of the testing takes place in close proximity to sanitary sewage systems, so it's not a job for the squeamish. Even in the foulest weather you have to be able to take accurate samples and secure them from the elements so they can be tested in the lab. It's also essential to have some knowledge of the industries you are investigating so you know exactly what you're looking for in the samples you take.

Yvette Kimber *is a water pollution investigator.*

How did you get the job?

Yvette Kimber has a degree in chemistry. She was working as an environmental chemist for a large oil company when she was laid off during a business reorganization. She decided to refocus her career by going for a master's in environmental science. Since she needed a job to pay for this bit of higher education, she checked in at the state employment office near the campus and found a position with a regional sewer district. The sewer district needed pollution investigators with college degrees and Yvette fit the bill.

What do you do all day?

Yvette will be the first person to tell you that the job sounds more glamorous than it is. Sure, she's watching for pollution violators and reporting the bad guys to her bosses at the regional sewer district in the Midwest,

> **CHECK LOCAL EMPLOYMENT AGENCIES' JOB BULLETIN BOARDS. VISITS TO PERSONNEL DEPARTMENTS IN MUNICIPAL, COUNTY, AND STATE GOVERNMENTS CAN ALSO PROVE HELPFUL.**

but that involves some very physically demanding and often dirty work. Every day Yvette and her partners go out into the field to check different companies and verify that their discharges into the environment or into the sewer system are within state and federal guidelines. The dress for this job is coveralls, blue uniform shirt and pants, and steel-toed boots. "In the winter it's long johns too!" Yvette adds. "We use a 28-bottle sampler with an automatic timer on it. Usually the discharge we're checking is separate from the sanitary waste, but occasionally we have to take samples where the operations discharge is combined with sanitary waste, and that is a dirty, disgusting job. When we've had to take samples from a street sewer," she says, "we sometimes get toilet paper wrapped around the sampler." You need a strong stomach.

The samples are brought back to the lab, logged in, refrigerated, and then analyzed. The data from those tests has to then be compiled in sometimes lengthy written reports.

Where do you see this job leading you?

Yvette believes the job is a good stepping-stone to other jobs in the environmental field, including everything from managing plants, to sampling work, to even designing plants. She wants to finish her master's degree and "get into waste-treatment operations," and perhaps, she says, even design plants.

26. Water Conservation Consultant

description: A water conservation consultant helps public water systems save water. Although many municipal water systems and private water companies have their own conservation departments, it is often necessary for them to reach out to private consultants to develop new plans and programs. These consultants examine supply and demand patterns, and devise strategies to save water on a long-term basis.

A conservation consultant will usually begin by researching supply and consumption statistics. Then he or she will investigate the system infrastructure—from the reservoirs or wells to the faucets and pipes that bring the water to the consumer. The consultant's advice can range from suggesting a maintenance plan to reduce leakage, to pushing for legislative changes that would govern certain types of water usage.

salary: Entry-level for someone with an undergraduate degree is $25,000. With a graduate degree and a couple of years of experience, the job can pay anywhere from $35,000 to $85,000 per year.

prospects: Excellent, especially overseas. The trend toward water-demand management appears to be a universal outgrowth of the population explosion. Wherever water providers are wrestling with the problem of increased demand and limited supply, conservation consultants are in strong demand.

qualifications: A graduate degree is highly recommended, although positions for those with undergraduate degrees are available. The educational discipline can vary widely; some people enter this field with technical backgrounds in engineering or planning, while others have majored in hydrology, geology, political science, economics, or accounting. Consultants trained in marketing, in fact, are in special demand, as water utilities try to develop strategies for changing the public's consumption habits.

characteristics: You need a combination of skills in this job. First, you have to have a technical or analytical grounding so that you'll be able to understand the problems. It's also essential, though, to have a grasp of human dynamics and even politics as you attempt to implement your ideas. You also have to be able to deal with continuing frustrations, particularly the politics associated with local and state management of water resources.

Amy Vickers *is a nationally known water conservation consultant working out of Amherst, Massachusetts.*

How did you get the job?

Amy Vickers graduated with a degree in philosophy and landed a job right out of college at the New York City Department of Environmental Protection. She took to environmental work right away, parlaying her experience in that first job to a position with the New York City Council's Committee on Environmental Protection. Following this, Amy acquired a master's degree in engineering from Dartmouth College.

She later began working for the Massachusetts Water Resources Authority as a project manager in the capital water engineering division. Eventually, she landed a job with a private consulting firm and managed to fit in postgraduate studies in geology and engineering. When company cutbacks made her subject to relocation, she decided to start her own company.

What do you do all day?

A project usually begins with Amy meeting with everyone involved in the water service project—from a town's mayor to the people who install meters. After initial evaluations, the real work begins: combing over hundreds of blueprints of a water system's infrastructure and identifying where water is being lost and wasted. In many cases, water is being lost because of leakage; this results in pipes needing to be replaced or retrofitted, or an entire system's distribution or treatment system needing to be revamped.

Amy has a network of technical experts who help in examining blueprints, sounding pipes, and checking meters for accuracy. She may make recommendations to increase efficiency in commercial and industrial processes, or may offer suggestions to change water use policy.

> THERE ARE SEVERAL ASSOCIATIONS TO JOIN, SUCH AS THE AMERICAN WATER WORKS ASSOCIATION, THAT MIGHT OFFER CONTACTS AND YEARLY CONFERENCES.

To calculate how much water will be saved, Amy relies on various computer modeling programs that she herself has developed. She identifies costs and benefits, and presents that information to the client. Projects are long-term and often take several years to complete.

In between all this work, Amy is very active in recommending and helping to write water conservation legislation, as well as in helping to lobby for these changes. She's also busy writing articles and is in demand as a speaker at many conferences and seminars.

Where do you see this job leading you?

The demand for Amy's expertise is growing worldwide, and she sees more opportunity for this work opening up in international markets.

27. Hydrologist

description: Hydrologists examine the movement of surface water and groundwater. They assess water resources, looking at quality and quantity by collecting samples and by measuring resources like rivers, streams, lakes, reservoirs, estuaries, and groundwater.

Their calculations and observations are then fed into computer models.

A young hydrologist can expect to spend half of his or her time outdoors doing field study work and the other half in the office doing analysis and writing reports. Writing the reports is often the most important part of the job.

salary: Entry-level hydrologists earn around $30,000 per year.

prospects: The job market for hydrologists is relatively good. A chief hydrologist with the U.S. Geological Survey says that there is a lot of opportunity in eastern Europe. He is also encouraging about job prospects in the United States, saying that the increased pressure to insure water quality will open up the field even more.

The U.S. Geological Survey is the largest employer of hydrologists in the country, employing about 2,000 hydrologists. Many jobs can also be found with federal agencies like the Bureau of Reclamation, the Army Corps of Engineers, and the Environmental Protection Agency, and with state governments. Private consulting firms are the second-largest employers of hydrologists.

qualifications: A bachelor's degree is not necessary to become a field technician.

To be a full-fledged hydrologist, a college degree is required. Most hydrologists, in fact, have Ph.D.s or master's degrees.

characteristics: One needs to be intellectually curious. A hydrologist also must be able to translate and communicate technical concepts to the general public. Writing and computer skills are essential, as is an interest in physics, chemistry, and math.

Pixie Hamilton *is a hydrologist and supervisor for the U.S. Geological Survey.*

How did you get the job?

Persistence pays off. It seems the federal government is under a perpetual hiring freeze, but don't let that discourage you. Pixie Hamilton didn't. Upon graduating with a master's degree in civil engineering and bachelor's degrees in civil engineering and environmental science with an emphasis in hydrology, Pixie visited the local office of the U.S. Geological Survey. The prospects were grim, but Pixie didn't let that discourage her. She kept at them, and within a few months was working there. She advises others to do the same. When it comes to the federal government, she says, the hiring status can change from one day to the next. "When a window of opportunity opens up, you want to have your paperwork ready."

> DON'T IGNORE LARGE ENGINEERING FIRMS JUST BECAUSE THEIR NAMES DON'T SOUND LIKE THEY HIRE HYDROLOGISTS. MANY DO—EVEN FIRMS THAT DON'T IMPLY EARTH SCIENCE OR HYDROLOGY IN THEIR COMPANY TITLES.

What do you do all day?

Pixie's first days on the job involved being out in the field on various projects taking measurements and collecting samples. The work included sampling water wells, ponds, and streams. She also looked at sediments, their composition, and how fast they traveled.

Today, she's moved up to the job of supervisor and works in the office much of the time, though when she first started the job she often found herself doing more fieldwork. Sometimes, the research required standing in the middle of a storm gathering data, which meant working nights and weekends.

Typical projects include examining saltwater intrusion into a coastal town's groundwater. That groundwater is the town's drinking-water supply. Another project involves the Chesapeake Bay, where a team of hydrologists is studying four major tributaries to the bay, assessing levels of nitrates and phosphorus. Her office also does a lot of work for military bases, drilling wells and drawing samples to analyze levels of contamination.

There's also quite a bit of computer work involved in being a hydrologist, since all of the data must be carefully analyzed. But Pixie says the most important part of the job may be translating the analyzed data into reports. These reports are the critical end result of each project and must be well written and easily understood.

Where do you see this job leading you?

Pixie is a supervisor now and wants to stay on the management track. As a manager, she now does marketing to solicit new clients. She also oversees projects and a staff of 45 employees.

28. Forester/Forest Ranger

description: Those working for such government agencies as the U.S. Forest Service, National Park Service, Bureau of Land Management, Fish and Wildlife Service, U.S. Army Corps of Engineers, and Environmental Protection Agency can best be described as protectors or caretakers of the forest. They are responsible for the planting, nurturing, and harvesting of forests. They are also responsible for improving and protecting wildlife habitat and water quality and for providing recreational access and facilities. An entry-level forester's work may entail working with planning issues regarding major development, creating trails for public access, doing research, or helping to battle a raging forest fire. As he or she moves up the career ladder, the forester moves from working mostly outdoors to working indoors. Promotion usually means more administrative, supervisory, and managerial duties, and more paperwork.

salary: You're not going to get rich being a forester, but you can make a good living. Salaries for foresters with the federal government start at anywhere from $15,000 to $18,000 a year, even for those with a master's degree. The average salary, however, is in the $30,000 range, while a forest supervisor can make in the mid-$60,000s.

prospects: Job prospects with the above-mentioned government agencies are very limited. Because federal budget cutbacks have forced downsizing, agencies simply aren't hiring. But there is a chance for you to break in, as hiring cycles change with the political climate.

qualifications: A bachelor's degree in one of the natural sciences is required. You need a strong fundamental understanding of natural resources. Backgrounds in engineering, geology, or computer science are also useful.

characteristics: A forester needs to be a critical thinker, articulate and able to work with people from all different backgrounds. Being a good negotiator helps, as a district forest ranger can spend a lot of time listening and responding to citizens' concerns and complaints. When a forest ecosystem comes under fire, you may be called upon to testify at city council or community board meetings, or you may find yourself in the middle of a battle between developers and environmental groups.

Corey Wong *is a forester for the U.S. Forest Service.*

How did you get the job?

While Corey Wong was in college, he got a job with a work-study program for the U.S. Forest Service assisting researchers in evaluating insecticides. He continued working for the service as a graduate student. Corey has a degree in forestry and a graduate degree in forest planning. He considers himself lucky to have gotten the job when he did. "We haven't had many entry-level people in the last few years," he says. He worked in several district offices before being promoted to district ranger. He's now in charge of a region covering 200,000 acres and supervises a staff of 15 to 30 people, depending on the season.

Following Corey's path through a work-study program is a good way to get into the Forest Service. "Now we hire specialists like wildlife biologists, geologists, archaeologists. Our newest person is a wildlife biologist from a college co-op education program. She's going to college and working for us at the same time. When she's done

we guarantee her a job. She's the only new hire to come through here in the last three or four years," Corey says.

What do you do all day?

Corey's office window looks out onto a hillside of trees and a spectacular view of the mountains. It's one of the perks of the job. Don't think for a minute that a forest ranger's job is mostly done while riding horseback; it's more managerial than anything else. He or she is responsible for all the activity on the grounds, for maintaining contact with community leaders, and for ensuring overall protection and maintenance of the forest. On any given day, a district ranger may oversee the harvesting of timber or the research and implementation of a controlled burn for a designated area. As the scenery changes with the seasons, so does the job. Managing a forest involves much more than concern for the trees.

Recently, Corey was checking on the construction and repair of trails and was about to head to a mountain summit to

> **TRY TO FIND A FORESTRY SCHOOL THAT OFFERS WORK-STUDY PROGRAMS. THESE CAN BE A WAY TO GET YOUR FOOT IN THE DOOR OF A GOVERNMENT FORESTRY JOB.**

check on the construction of a new toilet. On his way, he stopped to chat with highway employees installing new poles, as he was concerned with the color of the poles. "Later, I'm meeting with a man who owns a lot of property surrounded by national forest land, and we have concerns we need to talk about, including the problem that people are using it as a garbage dumping site." During winter months, he supervises the ski areas within his district. Within the forest there are many privately owned parcels of land; Corey oversees property surveys to make sure development is not occurring on protected forest land.

He is also dedicated to encouraging and ensuring a creative and prosperous work environment. Most of the area he's in charge of is accessible by roadway, although he frequently needs to hike to check a specific site, an activity he enjoys. He also gets to experience the treat of riding horseback from time to time.

Where do you see this job leading you?

"I plan on staying with the Forest Service for a variety of reasons. One is that I love what I'm doing. It's quite a challenge and I like working with the community," Corey says.

His job is certainly unique. Corey says that while other specialists, like wildlife biologists, can find comparable jobs in the private sector, he doesn't feel his mid-level forest manager skills are as easily transferable. Right now he loves the fact that he and his family are able to live in a small mountain town yet be less than an hour from a major city. "It's a great place to raise kids," Corey says.

29. Forester
(FOREST PRODUCTS COMPANY)

description: In the simplest definition of the job, a forester is responsible for the cultivation of forests. This includes the nurturing, harvesting, management, and conservation of trees. But the job covers much more than that. A forester deals not only with trees but also with the conflicting concerns about the proper use of forests, including improvement of habitat for wildlife versus recreational uses. Working for a forest products company, a forester deals with everyone from top company executives and government regulators to the loggers themselves. Not only does a forester select the trees to be harvested, he or she also hires logging crews and designs the entire tract, including the layout of roads needed to do the work.

salary: For private forest products companies, you can earn a starting salary of $25,000 to $30,000 per year. With 10 or more years of experience, you can earn up to $70,000 as an upper-management regional forester.

prospects: Competition is stiff for positions in private companies. The jobs vary according to region of the country. Many of the forests in the West come under the guidance of federal agencies, while in the Midwest and East you may find opportunities as a consultant for private landowners or timber companies.

qualifications: A forestry school degree is a must. This position requires a background in natural sciences as well as statistical procedures and computer science. A knowledge of engineering, geology, and other sciences can give you an edge.

characteristics: People often have the romantic misconception that foresters simply ride around on a horse alone throughout the woods. Actually, foresters need people skills. They must be articulate, diplomatic, well educated, and informed. They also need to have the motivation to continue self-education about changing industry methods and techniques. They must be able to tolerate working in all weather conditions—it can get pretty cold in mountain snow.

Robert Bass *is a forester for a privately owned forest products company.*

How did you get the job?

While in college earning a bachelor's degree in forestry resource and management, Robert Bass worked summers for the U.S. Forest Service and private forest products companies. After graduation, he got a forester job working for the university's forest. From there he worked as a forester for an Indian tribe, before returning to the university's forest. Each position provided an increase in responsibility and compensation. Today, Robert is a forester for one of the country's largest forest products companies.

What do you do all day?

Every day is different. One day he may be flying in a helicopter surveying some of the 180,000 acres that he and 11 other

> FORESTRY IS NOT SOMETHING SIMPLY LEARNED IN THE CLASSROOM. GETTING SEASONAL WORK WHILE IN COLLEGE IS VITAL. NOT EVERYONE HAS THE DRIVE AND WILL TO TAKE SOME OF THE LUMPS OR LOW PAY THAT COME WITH OFTEN-FRUSTRATING SEASONAL OR TEMPORARY JOBS.

foresters manage. Many days he's on site, checking with loggers and other field crews he's hired. Other days he may spend in the office doing all the legally required permit paperwork. When he started the job a year ago, the first thing he did was set up a rotation plan for all the tracts. Once he selects the tracts to be planted, cultivated, or harvested, Robert does the mapping and verifying of all boundaries. He also designs the roads to move equipment in and out of selected areas. As a forester, he also troubleshoots all the problems that pop up. For example, if he sees that bears are eating the trees (he says they love to eat the sugar layer beneath the bark of the most valuable tree species), he puts out feeders—barrels of food to keep the bears from eating up the forest.

Robert says his work has the greatest deadlines of all: the seasons. "If you miss those windows of opportunity, then you miss them for the entire year. For example, if you spray too early, it'll have no effect at all and you've wasted $100,000 to $200,000," he says.

Robert loves the flexibility and independence of the job and, of course, the breathtaking scenery.

Where do you see this job leading you?

There's plenty of opportunity to advance up the career ladder within the company. From senior forester Robert can move to area or district manager. The only downside is that the further up you move, the more you are stuck working indoors rather than outside amid the trees.

30. Forest Archaeologist

description: Forest archaeologists do spend a lot of time outdoors, but they don't spend very much time digging. These archaeologists are working to make sure that no historical or cultural resources are damaged by the activities going on in our national forests and parks. Before a campsite can be extended, an area chiseled by miners, or trees cut by lumberjacks, a forest archaeologist must first make sure that no significant resources will be destroyed in the process.

Going out and looking at an area that lumberjacks or miners want to go into is a big part of this kind of archaeologist's job. But that is only half of the job. Once a cultural resource has been identified, it must be preserved in such a way that it can be used for the enjoyment or education of the general public. Forest archaeologists often find themselves organizing or helping to create a way for these cultural resources to be displayed.

salary: A person with a bachelor's degree could start out as a forest archaeologist trainee making $20,000 a year. In order to advance beyond this point, it is necessary to have further education. A person who has begun graduate work will still carry the title of trainee, but will be able to take on more responsibilities and bring in a salary of around $24,000. After about two years into his or her graduate studies, a trainee can become a forest archaeologist and bring in $35,000 a year. This salary can increase to as much as $45,000 to $60,000.

prospects: While the job market is full right now, it can be expected to thin out within the next 5 to 10 years. The Archeological Resources Protection Act, which was passed in 1974 and once again in 1979, required that the federal government hire a steady number of archaeologists. This bulge in the workforce is expected to begin thinning out, creating a number of jobs along the way. Those starting their bachelor's studies within the next year or so should be in a good position to find work in this field by the time they graduate.

qualifications: A person wanting to get into this field should seek a bachelor's degree in anthropology. But in order to advance in this line of work, some sort of graduate work must be completed, either a master's or a doctorate.

characteristics: Archaeologists have to be dedicated to their jobs. The initial attraction for most archaeologists is almost always fieldwork. They usually enjoy working by themselves but also are able to work with others because of experience working with a crew in the beginning stages of their career.

Will Reed *is a forest archaeologist for a national forest in Idaho.*

How did you get the job?

Will Reed first became interested in archaeology at the age of seven, when he began reading everything he could on the subject. Throughout his high school years, Will continued to learn more about archaeology through extracurricular activities, but it wasn't until after he served for two years in the U.S. Air Force and then graduated from college that he was able to start his career in archaeology. Equipped with his bachelor's in anthropology, Will got a job working for the Forest Service.

However, two years after Will took that job, the federal government began requiring all archaeologists working for it to have a graduate degree. Will then began working in the field as a private contractor. At the same time he went back to graduate school to fulfill the requirement that would allow him to work in a national forest again.

Following graduate school, Will worked at Idaho State University, where he organized a contract laboratory in the anthropology department, before he was brought on as a forest archaeologist at Boise National Forest. "Somewhere along the line, I got the idea that history does have a lesson to teach and I began to feel that it was important to maintain some of the lessons that history has to offer," he says.

What do you do all day?

Because the Forest Service encompasses so many different functions, Will spends most of his time monitoring the various goings-on. "There are a lot of different activities that are primary activities of the Forest Service, and they all require some input from the archaeologist in order to make sure that they are not hurting cultural resources. We try to address the needs of recreation, of miners, cattle, of ranchers."

On an average day, Will might start by heading out with a group from the mineral shop. If there is a particular area that they are wanting to mine in, he will determine whether or not there are any resources in that area that need to be protected.

If there are, he will suggest a way that the miners could work and not affect the resources. He may go out to view three different sites with proposed activity in one day. He will also usually spend a large majority of his day working on projects that enhance the educational use of these cultural resources for the public. For example, Will is currently working on a heritage trail that takes hikers through an area once mined by Chinese immigrants, which will feature interpretive signs explaining what the landscape is all about and the role of the Chinese miners.

Where do you see this job leading you?

From here, Will could stay on the government track, which would eventually put him in more of an administrative role, or he could move into the private sector as a consultant or contractor.

THE BEST RESOURCE FOR FINDING JOBS IN THIS FIELD WOULD BE THE NATIONAL WILDLIFE FEDERATION'S <u>CONSERVATION DIRECTORY</u>. FOR INFORMATION ON THIS DIRECTORY, WHICH LISTS ALL WILDLIFE ORGANIZATIONS IN THE UNITED STATES, CALL OR WRITE THE NATIONAL WILDLIFE FEDERATION. (FOR ADDITIONAL INFORMATION, SEE RESOURCES, PAGE 212.)

31. Park Ranger/Nature Interpreter

(FEDERAL GOVERNMENT—NATIONAL PARK SERVICE)

description: They used to be called guides, but the National Park Service increasingly refers to many of its park rangers as interpreters—people whose primary job is to help visitors understand and appreciate what they experience at various national parks. Because human interaction with the park's environment also creates safety and conservation issues, even rangers who primarily work in this "public relations" role need to perform some law enforcement and emergency functions as well.

salary: Salaries range from $20,000 to $30,000 a year. Compensatory time is offered for extra hours worked, and subsidized on-site housing is available.

prospects: There is tough competition for permanent positions. As the federal government downsizes, employees from regional offices are moving back into the field, making entry-level access more difficult. Many interpreters start as part-time seasonal rangers. Some colleges offer co-op programs in which students first work for the Park Service for credit, and then apply for permanent positions.

qualifications: A bachelor's degree, preferably in the environmental field, is desirable. Seasonal or volunteer work is a good way to get your foot in the door. The Park Service requires rangers to attend on-the-job training workshops.

characteristics: An interpreter has to know practically everything about the park and be able to inform visitors about what they're seeing. You have to be absorbed in the park surroundings and be eager to research and learn about the indigenous wildlife, woodlands, ponds and streams, and distinctive characteristics which make that particular park unique. As an interpreter you are both a teacher and a protector.

Sue Branch *is a park ranger at Yosemite National Park.*

How did you get the job?

As a kid growing up, Sue Branch and her family often vacationed at national parks. She loved those trips so much that, at age 16, she wrote in a career questionnaire that she would one day like to be a park ranger. Her interests changed, however, and she went on to college to earn a bachelor's degree in biology. Unsure of what to do next, she remembered her teenage interest in being a park ranger, and soon began graduate work in natural resources interpretation. As part of that program, she took an internship as a seasonal ranger at Yosemite National Park. For the following two summers, she was hired as a seasonal ranger, and eventually was hired on a permanent basis. Her first role as ranger was as a nature interpreter. Sue then decided to move into the education department, where she now works as the department's coordinator.

In her spare time, she also went back to school and earned teaching certification.

What do you do all day?

The role of a park ranger interpreter is to educate the visitors about the park, along with all the wonder, beauty, and resources it offers. This is done by taking visitors on guided walks and talks. The day may be filled with ushering a group of visitors through one particular area of the park, or staffing a public information booth. It may also mean narrating a slide show at one of the pavilions.

After working as a park ranger interpreter, Sue decided to specialize in the area of student education. As the education coordinator, it's her job to oversee and manage several programs for elementary school students. She does everything from scheduling class trips to writing lesson plans. In some cases, Sue will send lesson plans to teachers for use in the classroom, in advance of a field trip to the park. At the park, Sue conducts special seminars for the students, and leads them on guided nature walks. As an administrator, Sue oversees several other park ranger educators. She also writes the curriculum for these programs, and coordinates a junior rangers program—a special course for junior high school students.

There have been emergencies where Sue and other interpretive park rangers we are called to help out in unusual situations. Two years ago, as firefighters battled raging fires in Yosemite, Sue helped in escorting the media to the fire lines, answering reporters' questions, and making sure they didn't tread into dangerous areas. Other rangers with specialized training helped in evacuations and in search and rescue.

Working in such a glorious environment amid picturesque waterfalls, lush forests, and rolling rivers is one of the best things about the job, she says, but it is meeting people from all over the country, and from around the world, that is the most rewarding aspect of this career for her.

Where do you see this job leading you?

Sue loves the educational work that she's doing, and plans to stay with the Park Service. Eventually she'd like to work in a smaller national park that offers a more intimate setting.

> **SINCE IT IS EXTREMELY DIFFICULT TO GET A JOB WITH THE NATIONAL PARK SERVICE, ONE SUGGESTION IS TO TRY TO GET A POSITION WITH ANOTHER GOVERNMENT AGENCY THAT OFFERS OUTDOOR WORK.**

32. Law Enforcement Park Ranger

description: Though it's hard to think of a park ranger without envisioning Smokey Bear putting out forest fires, this is only one of the tasks these hardworking individuals must perform. Patrolling the parks either by horseback or on snow skis, conducting search and rescues, and making arrests for illegal activity in the parks are also among the duties a law enforcement ranger can be expected to be assigned. The work of a ranger is physically rigorous and almost never 9 to 5. Shifts can go on for more than 24 hours if the ranger is assigned to backcountry trail management or to a search that goes on for days. But what makes this work rewarding is that everything a ranger does is geared toward protecting the natural resources within the park and providing the public with a safe atmosphere in which they can observe and enjoy these resources.

salary: A full-time, entry-level position starts at about $29,000 a year. A law enforcement ranger with about 10 years of experience could expect to make at least $35,000. Seasonal workers are paid hourly wages; the current rate is approximately $9 an hour.

prospects: With the continuing trend of downsizing in the federal government, positions for full-time law enforcement rangers do not often become available. But these jobs do exist, and the best way to find them is to take a seasonal position and keep your eye out for a full-time opening.

qualifications: A bachelor's degree in wildlife biology or environmental science is suggested for anyone interested in gaining employment as a ranger. Seasonal law enforcement rangers must also have a seasonal commission, which can be obtained from any college that offers the training program. In addition, not all training is obtained on the job. Law enforcement and emergency medical technician training is required before applying for work as a seasonal law enforcement park ranger. Refresher courses are provided by the Park Service. Most rangers have worked in parks in a volunteer or seasonal capacity before going through the training.

characteristics: Law enforcement park rangers must be outgoing and able to get along with people they encounter on the trail. It is also important that they be independent enough to work on their own and confident that they can handle situations by themselves. These rangers also must be physically strong enough to handle the often challenging situations in which they will find themselves from day to day.

Joe Evans *is the chief park ranger at Yosemite National Park.*

How did you get the job?

Joe Evans has a bachelor's degree in American history. Like most other national park rangers, Joe began his career with the Park Service working as a seasonal employee. After five years working in that capacity, Joe was hired on a permanent basis. He's worked in several national parks, including the Grand Canyon, Yellowstone National Park, and others in Arizona and Hawaii. He moved up the ranks through various supervisory positions, and was eventually promoted to one of the park's top jobs, that of chief park ranger. He's been a park ranger for more than 20 years.

As a ranger, Joe started out doing wildlife/nature interpreting work, but eventually switched to doing the law enforcement work.

What do you do all day?

As chief ranger for the park, much of Joe's time is spent taking care of administrative and personnel responsibilities. He's in charge of 17 permanent rangers and 80–90 summer seasonal rangers. The park rangers are involved with many duties, including campground management, entrance stations, visitor services, law enforcement, and search and rescue efforts.

"Your primary responsibility is to protect the resources of the park, and also provide for visitor safety," says Joe.

The job responsibilities can vary from one national park to another. Some parks, he says, have different law enforcement needs than others. For example, in some parks the rangers' time is taken up with managing visitors and concession employees; in others, the rangers' work focuses more on resource protection, such as special watches during hunting seasons, or patrols to control illegal activities.

Search and rescue efforts are a big part of the job. "Just last night we had a major search and rescue. A rock climber fell and broke his femur, and it took us six hours to get him off the cliff," Joe says.

You also never know who you'll encounter. Joe says that, in making routine traffic stops, they've pulled over motorists who turned out to be wanted felons. Just last summer, he says, they stopped a motorist who turned out to be wanted on homicide charges.

Where do you see this job leading you?

Joe says he'd like one day to be a superintendent (top administrator) of a national park.

> **GETTING A JOB WITH THE NATIONAL PARK SERVICE IS VERY DIFFICULT, SO YOU MIGHT WANT TO TRY YOUR LOCAL CITY OR STATE PARK SERVICE.**

33. Parks Planner

description: Parks planners usually work for city, county, or state agencies. The job of a parks planner is to provide for the optimum use of public parks. That includes creating the long-range plans, designs, and recreational features for the ever-changing demands of the public. This must be accomplished while keeping the environmental integrity in balance.

On any given day, the parks planner may have to design a playground or begin to research a 20-year plan for an entire city's parks system. All projects begin with research, and then it's on to the drafting board. A good parks planner is creative and able to satisfy the community's demands within the constraints of a budget. You may be coordinating construction of a picnic area in one park and designing new hiking or mountain bike trails in another, while writing rules to keep in-line skaters off walking trails.

salary: Entry-level salaries can start at $15,000 a year, and after several years of experience may reach $35,000.

prospects: Parks planning may be a tough profession to break into and openings may be somewhat limited.

qualifications: Those intending to be parks planners should get experience in landscape architecture or recreational planning. These backgrounds, coupled with a degree in public administration, can put you on the administrative track.

characteristics: Vision and creativity are the two most important traits needed for this career. But one also needs to possess a willingness to compromise under budgetary restrictions. The parks planner also needs to be receptive and open-minded to listen to community opinions, needs, and desires.

Douglas McRainey *is a parks planner who works in Cleveland, Ohio.*

How did you get the job?

Douglas McRainey started out wanting to be a lawyer but realized his interest was really in landscaping. In graduate school, he got a job with the Ohio Department of Resources working on a project to increase recreation areas near the Ohio River. He saw a notice for his current job posted on the bulletin board and was hired specifically to work on the Ohio Metroparks master plan. This 20-year plan will serve as the blueprint and policy for parks in three Ohio counties. For this master plan, he gathers and analyzes research data, draws schematics, and offers recommendations for the approval of the board of directors and the community council.

What do you do all day?

"Every day is different. I'm in the office half of the time and actually outdoors in the parks the other half. I spend a lot of time meeting with park managers and community groups."

It's busier in the spring and summer, when the weather is nicest. That's when Douglas spends most of his time in the parks checking on future projects and ongoing construction. In the fall and winter, he spends more time in the office. "The first thing I do is go through the mail, schedule meetings, and make phone calls." He's con-

> A KNOWLEDGE OF GEOGRAPHICAL INFORMATION SYSTEM, A TYPE OF COMPUTER SOFTWARE COMMONLY KNOWN AS GIS, WILL GIVE YOU AN EDGE.

stantly in touch with other city and county agencies, coordinating plans. Douglas works on several projects, and delegates some of that work to interns.

On a typical day, Douglas works on his major project, developing a GIS (Geographical Information System) program for the park district. The GIS will enable park department staff to analyze resources and other data. While he's doing that, his boss may interrupt and request a site plan for a specific park. Once the GIS is completed, he has to write and send out a memo regarding problems with the computer software they're using. After lunch, he may take a second look at a plan to relocate picnic areas in one park, while in another park he may decide to remove the picnic areas to create a wildlife management zone.

The job's challenge is that the public's demands are con-

stantly changing. "Five years ago, who would have thought that we'd have to make allowances for Rollerbladers? Then there are the walkers and joggers who complain about the mountain bikers using their trails. We also have mountain bikers who want to get off onto more rugged terrain. Who knows what it'll be 10 years from now?"

Where do you see this job leading you?

Since Douglas is a senior planner, the next step up is administrator. He'd like to make the advance and manage a planning office. If he were to move to the private sector, it would have to be for a consulting firm that works in recreation. He could easily make that transition, given his experience and extensive GIS knowledge.

34. Soil Scientist

description: There is a broad range of careers in the area of soil and water conservation and remediation. The jobs of agronomist, forester, geologist, and civil engineer all come under the heading of natural resource careers, but there are also specialists such as soil scientists, soil technicians, and soil conservationists. Soil scientists examine the properties of the terrain to determine land use or remediation. Their calculations measure erosion and basic soil qualities such as chemical content, moisture, aridity, and rock content. They're often an integral part of engineering surveys and land-use designs, and of conservation plans and practices. Their jobs vary with their areas of expertise. Some soil specialists determine toxins dumped at sites, while others may be hired to help farmers yield more productive crops.

salary: According to the Soil and Water Conservation Society, soil scientists start at a salary of around $18,000 a year, and those with a master's degree start at around $24,000. After about 10 years of experience, a soil scientist may earn $40,000 to $50,000. Promotion into management can boost salaries to around $65,000.

prospects: Many soil scientists are employed by government agencies, but more are being hired by private consulting firms and in the chemical, oil, and gas industries.

qualifications: The level of expertise demanded in this field requires a strong background in the natural sciences. A degree in a related field such as agronomy, forestry, biology, or agriculture with a concentration in soil science is also a good way to go. You'll need at least a bachelor's degree to get a good job in this field; a master's degree opens up even more doors.

characteristics: The traits that will set you apart from the pack are good communication and writing skills. Soil scientists find that they need not only technical knowledge but also the ability to communicate their findings in written reports and articles.

Myra Peak *works as a soil scientist.*

How did you get the job?

Straight out of college, Myra Peak headed for work in the coal mines. It wasn't a typical job for a woman, but with degrees in physical science and mathematics, she got work as a soil sampler. Her responsibilities included surveying land parcels, creating maps, and doing lab work. She worked her way up through the ranks and was promoted from soil scientist to reclamation foreman, where she supervised crews as that company's first female foreperson. Myra eventually went back to school to earn a master's degree in soil chemistry.

After five years in the mining industry, Myra switched gears a bit and went to work in the hazardous waste industry as a remediation contractor. Making the leap from the mining industry to hazardous waste took a bit of forcefulness in convincing an employer to hire her. "I walked in the door and said, 'You need me.' I took my resume and tried to show them how my skills would apply. I finally said to them, 'Put me in the field and make me do the job.'"

Eventually Myra decided to be her own boss by creating her own company, which investigates contaminated sites. Myra does the sampling, identifies content, quality, and sources of contamination, and offers remediation recommendations. Her staff of seven employees includes a hydrogeologist, a field biologist, and field technicians.

What do you do all day?

Myra spends many days outdoors on site. "I'm literally out there digging by hand taking samples," she says.

As owner of her company, Myra splits her time between outdoor fieldwork and office work, such as writing reports, examining lab results, formulating evaluations, and performing company administrative duties. The size of the project sites she works on can range from a small car wash to a 20,000-acre ranch. She and her staff try to determine the likely places people might have used chemicals and

> THE FEDERAL GOVERNMENT OFFERS PROGRAMS FOR STUDENTS. CONSIDER VOLUNTEERING TIME AT A SOIL AND WATER CONSERVATION AGENCY IN YOUR AREA. GAINING EXPERIENCE WHILE IN SCHOOL WILL GIVE YOU A HEAD START.

uncover potential sources of contamination, such as abandoned mines or oil wells, treatment operations, or a fire in a building where transformers contain PCBs.

The job involves some detective work, as the soil scientists must determine where to draw samples and must know what they're looking for. For example, one project involves a mine where a petroleum storage facility has had numerous leaks. Despite a "cleanup," contamination is still detected. Myra has

been called in to find the source of the pollutants. "We are literally letting our noses tell us where we have to go. We are watching them dig. They put it [dirt] in a dump truck and drive it to a landfill, spread it out and aerate it to allow the petroleum hydrocarbons to volatilize." It'll then be sampled, she explains. Before excavation, they took samples, interpreted findings, and extrapolated on a map how far the contamination had traveled.

Myra's advice to others interested in this work is to get a strong scientific background. She says her master's degree in chemistry has proven to be invaluable.

Where do you see this job leading you?

Successfully creating her own company in an industry with few women has been a great challenge. In fact, Myra's in demand as a speaker on both being a successful businesswoman and educating the public about soil science. She's also branching out, as she conducts training in the safe handling of hazardous waste.

35. Seismologist

description: As a seismologist, you will monitor the earth's subterranean movements in order to determine when earthquakes or aftershocks might occur in your region and how volatile they will be. Your research could also be used to investigate nuclear explosions or sonic booms, as well as for studies of the earth's interior, conducted for the purpose of ascertaining why the earth works the way it does.

Because the earth's interior movements can be monitored through computer-generated reports, this type of work is usually done in an office setting. You will spend a lot of time sitting in front of the computer and troubleshooting, as well as generating reports that will be of interest to many in both the academic and the public sectors. In order to draw conclusions about the earth's movements from the reports generated, numerous calculations must be made. This job can be tedious, and it requires a certain level of patience from the person disseminating all of the information. Generally, those in this type of work are able to stick with it because of their genuine interest in the physical nature of the earth.

salary: A staff seismologist can expect a salary upwards of $50,000 a year.

prospects: Recent cuts in the federal budget have pared the number of job opportunities available in the area of earthquake research. It is uncertain if this trend will continue in the future.

qualifications: Ideally, someone working in this position will have an academic background in geology with an emphasis on mathematics. It is also important to have an advanced degree, such as a doctorate. It is imperative that someone vying for this type of work be familiar with computers.

characteristics: People working in this field are generally inquisitive and detail-oriented. Seismologists must also have the patience to work on a statistical base that could take years to build.

Kate Hutton *is a seismologist at a technical institute in California.*

- -

How did you get the job?

Kate Hutton earned both her master's and doctorate degrees in astronomy, and planned to expand on her interest in the stars in her professional life. But after completing her post-doctoral work at Goddard Spaceflight Center, Kate was presented with a job market that was less than promising in the area of astronomy. "At that time NASA was downsizing and everybody was going into academia, so there weren't any academic positions either."

Because Kate's work at Goddard had been in the geophysics lab and involved a good grasp of computer skills, she found that her knowledge and skills were transferable to a field that focused on what was below her feet, rather than over her head, as the staff seismologist at California Institute of Technology.

What do you do all day?

Kate usually starts out her day by checking the computers to see what type of activity has occurred underground in the Southern California region during the night. The information comes to her over a microwave link system. Kate, with the help of about four assistants, must review all the reports, then select interesting information to be disseminated on the Internet. In addition to this, Kate is responsible for generating a weekly report that is also published on the Internet. All week she is compiling data for this report, but the bulk of the work, the writing, is done on Thursdays. Occasionally, Kate will spend time speaking to the media or documenting for insurance companies or employers that an earthquake did occur at a certain time and location.

When an earthquake occurs, everything changes. Kate remembers rushing into the office sometime around 4:30 on the morning of the Northridge earthquake. The first thing she did was make sure the computers were operating properly. Then she started trying to determine the size of the quake and the location. At first, she says, the information coming in was vague; this was because the computers were receiving so much data, they were running slower than usual. After she located the quake and determined its size, the rest of Kate's week was spent forecasting the occurrence of aftershocks and passing this information along to the media.

Although Kate admits her job is unpredictable, she says that that is one of the aspects she likes most about it. "I may have plans for my day, but nature sometimes has other plans," she quips.

Where do you see this job leading you?

Kate doesn't plan on any immediate career changes, but someone with her experience either could eventually move into an academic position, or find work in the private sector producing seismic risk reports and conducting disaster mitigation.

> **THE BEST PLACE TO LOOK FOR THIS TYPE OF WORK IS IN THE GEOPHYSICS DEPARTMENT AT YOUR LOCAL COLLEGE. THE CLASSIFIED SECTION OF THE GEO TIMES, A TRADE PUBLICATION, IS ALSO A GOOD WAY TO FIND OUT ABOUT OPENINGS.**
> - - - - - - - - - -

36. Nature Cartographer
(FEDERAL GOVERNMENT)

description: A cartographer makes maps. In wildlife management and environmental science, he or she is also called upon to analyze data so it can be distilled into charts and reports. The work deals not only with geography, but also with mapping and charting of wildlife movement, shoreline erosion, and species ranges. This is support work done primarily at a computer.

salary: Salaries start in the low $20,000s.

prospects: Nature cartography is a growing area for employment, especially with newer computer programs now available to translate the voluminous data collected by wildlife and environmental agencies. Right now, though, there is debate within the field as to whether this is a job for trained geographers or more of a computer science position.

qualifications: A degree in geography, geology, or a related field is required. Up-to-date computer knowledge is also a must—specifically GIS: Geographic Information System software, a basic computer tool used in related fields such as planning, forestry, and environmental consulting.

characteristics: You will need to be self-motivated, independent, and intuitive. You'll be taking raw information and applying it to get a desired product. A basic knowledge of environmental science is essential. This is an indoor job—you'd better love working at a computer.

Troy Mullins *is a cartographer/geographer who works in the Everglades National Park, Florida.*

How did you get the job?

Troy Mullins became fascinated with geography during his freshman year of college. As he zeroed in on his bachelor's in geography, a professor told him about an opening at the Park Service and helped to set up an internship that got him in the door there. He was soon hired as a staff member.

What do you do all day?

Recently, Troy was at his computer, working up a series of maps and charts on the movement of manatees. The people he works with want a chart of the population density within a certain part of the park and a map that shows the manatees' seasonal directions of travel. But Troy configures all types of maps. He takes direction from the researchers, drawing maps for whatever they request. After running the researchers' data through the computer, he can plot a map for their projects, be it anything from wildlife mapping to fire mapping to tracking the travels of manatees.

Troy says that what he enjoys most about the job is the rapidly changing computer technology. "There's a lot to learn, new software and tools to do the job." He says every project is different, so he has to almost constantly find new applications for his skills. It's one of the things he likes about the job, besides the regular hours.

Where do you see this job leading you?

Troy hopes to stay with the Park Service and perhaps get a promotion to manager, which would mean more responsibility and a higher salary.

> DON'T EVER SEND A RESUME ALONE; ALWAYS ACCOMPANY IT WITH A COVER LETTER DIRECTED TO A SPECIFIC PERSON.

37. Nearshore Oceanographer

description: A nearshore oceanographer conducts research on shorelines, looking at the processes responsible for erosion, as well as the accumulation and transportation of pollutants. In this line of work, you're most likely to be a government employee. The work is scientific research with an immediate, practical application. Since most Americans, for instance, live within 100 miles of a coastline, the information gathered is of intense interest to policy makers, businesspeople, property owners, and environmentalists. "It's not ivory tower research," one professional notes. "In some ways you're a slave with many masters." It is, however, useful, practical work with a vast number of applications.

salary: Entry-level salaries start in the low $20,000s per year.

prospects: In the larger field of oceanography, there's a shift in progress toward coastal research. Large oceanography-funding organizations realize that as you move into the coastal zone, you're beginning to address problems that have a direct impact on large segments of the population. Potential employers include universities, state geological surveys, state departments of natural resources, and county governments. Consulting firms are also hiring some oceanographers for engineering applications, as they try to design coastal construction projects that mitigate or reduce the effects of erosion.

qualifications: Most positions require a minimum of a master's degree. You will need a working knowledge of ecology, hydrology, meteorology, and geology. Scuba training is a must.

characteristics: Nearshore oceanographers need a sense of balance and perspective about their work, along with a genuine interest in the underlying problems being studied. They frequently find themselves straddling the worlds of pure research and the pragmatic concerns of people living and working in close proximity to the shoreline. They are also required to do a great deal of analysis to isolate problems from among a host of ever-changing variables. It helps to like the beach, although a lot of the work involves data processing and number crunching indoors.

John Haines *is an oceanographer with the U.S. Geological Survey's Coastal Center in St. Petersburg, Florida.*

How did you get the job?

John Haines focused his interest on coastal erosion while earning a bachelor's degree in environmental science. He went on to earn a Ph.D. in oceanography, and, although it isn't required, it certainly helped when it came time to look for work. There was also a bit of luck involved: "They were looking for someone when I was looking," John says. "It's a relatively small field and everybody knows everybody. I just heard about the job at professional meetings. I fit the bill, and there was probably no one else in the running with a Ph.D. in a relevant field."

What do you do all day?

It depends. Some projects involve periods of intensive fieldwork—often in the water, sometimes in scuba gear—as monitoring equipment is installed or checked. But oceanography has become so high-tech that John, by way of live remote cameras, can look at Ohio's Lake Erie shoreline from his office in Florida. As projects enter the monitoring stage, more time is spent indoors, perhaps studying information sent through satellites or through remote cameras installed along endangered coastal areas. There is also a constant process of mapmaking, to document the changing coastlines. On some projects, the bulk of John's time is spent at the computer, preparing reports, analyzing data, and making maps. But recently, he and members of his team were in the waters off North Carolina nearly every day collecting information about the waves and currents.

> YOUR RESUME IS ONE OF YOUR MOST IMPORTANT SALES TOOLS. IT MIRRORS YOUR PROFESSIONALISM NOT ONLY BY LISTING YOUR EXPERIENCE BUT ALSO BY ITS APPEARANCE. KEEP YOUR RESUME BRIEF AND UNCLUTTERED. IT SHOULD BE NO MORE THAN TWO PAGES LONG.

Where do you see this job leading you?

John is on the career path he envisioned he would be when he joined the U.S. Geological Survey. As his responsibilities grow, he admits that less and less of his time is spent at the seashores that first attracted him to the work. He does, however, enjoy bridging the gap between the scientific specialists doing basic research and the interest groups trying to address specific problems relevant to their shoreline. "By fitting in between," he says, "I satisfy my basic scientific curiosity and at the same time provide a service."

38. Geologist

(OIL AND GAS INDUSTRY)

description: Geology is the science of the earth—the study of the earth's origin, structure, and history. For many years, the biggest employers of geologists were the petroleum, oil, and gas industries. Unfortunately, the economy has changed that, and what was once an open market for geologists has now nearly shut its doors completely. There are, however, still a few hardy geologists making a living using their expertise to locate new sources of oil and gas.

Petroleum geologists are usually experts in a particular region. Their jobs entail poring over geological maps and studying an area's history of energy sources. They use technology, sampling, and computer modeling to determine where to strike it rich. The work is perfect for one who loves a combination of spending time out in the field, inside a laboratory, and in front of a computer.

salary: An oil company geologist with a master's degree can expect to start at $35,000 a year.

prospects: Unfortunately, the job prospects for petroleum geologists are not good. According to the Geological Society of America, many geologists in the petroleum industry have lost their jobs, and the situation does not look like it's going to change in the near future.

qualifications: A master's degree is generally required. Studies should include mineralogy, petrology, and sedimentation. Oral and written communications skills are extremely important.

characteristics: You've got to be curious—the petroleum geologist is in large part an explorer seeking out new resources. Having some artistic talent also helps—you should be able to draw maps in order to communicate to people where to drill. Since many companies no longer have geologists on staff, a lot of these scientists are working independently. A successful independent consultant needs to be able to hustle to create and manage business.

Harrison Townes *is a petroleum geologist.*

How did you get the job?

Harrison Townes started his career by working for oil companies. It has always been his job to find new rich oil fields or expand existing ones. However, for most of his career he's owned his own business. He's invested in and looked for oil for himself, but mostly is independently contracted by oil and gas companies.

What do you do all day?

Harrison is hired by companies to find oil and gas. It's called "wildcatting." Companies pay him to develop geological leads and provide prospects on where they should drill. He seeks out

> THERE ARE SEVERAL VERY HELPFUL GEOLOGICAL PROFESSIONAL ASSOCIATIONS THAT ADVERTISE EMPLOYMENT OPPORTUNITIES IN THEIR TRADE PUBLICATIONS.

new oil fields and checks out old ones. "Ideas come out of my head based on experience and knowledge of an area, and I make maps and get well data from wells that have been

drilled. I sniff out all sorts of leads and decide where there's a good place to drill for oil." Harrison is paid either a contracted amount of money or royalties.

At most, Harrison's in the office two to three hours a day. The rest of the time, he's in the geological library sifting through research, including well data, electrical surveys, other technical reports, and maps. He tries to uncover any leads or clues that he can.

When he's out in the field, Harrison travels with a microscope, sample trays, fluorescent lights, a pair of boots, a tin hat, and old clothes. Once he decides where to drill, Harrison patiently oversees the operation. This part is called "sitting on a well." That means staying out

for several days and nights supervising the drilling. "When you get to the depths of a zone that you think might be a pay zone, you need to be there to analyze the cuttings that come up from the bottom of the well . . . look at them through the microscope and determine whether you're in the formation with a show of oil or gas."

Where do you see this job leading you?

Harrison's been in the business more than 30 years and is still not ready to retire. He's made most of his money from investing in potential oil and gas prospects. Throughout his career, he says, he's made and lost a lot of money and keeps on hoping to hit it rich one more time.

39. Geologist–Economic (Mineral) Specialty

description: Economic geologists specialize in the study of mineral and metal deposits. The information they collect and analyze is used not only in the mining industry but also by federal and state agencies for a variety of land-use decisions in lumber production, recreation, wilderness studies, and conservation. This is a job for someone who loves the outdoors, although the work does involve laboratory analysis and office paperwork as well. The fieldwork includes the identification of rocks and the use of chemical analysis to learn the age and history of formations; this type of information can shed some light as to what's underneath the earth's surface, without having to drill a hole.

salary: Someone with a master's degree can start at around $30,000 a year. With a Ph.D., the starting salary ranges between $36,000 and $43,000.

prospects: According to the Geological Society of America, employment in the mining industry stabilized in the late 1980s but appears to be growing more active again. The U.S. Geological Survey is the largest employer of geologists in the nation, but jobs are also found at state geological surveys and at universities. Opportunities are opening up outside the United States as well, especially in South America.

qualifications: A master's or Ph.D. is preferred by most companies. The Geological Society of America recommends taking advanced mineral deposit courses, as well as courses in math, physics, geochemistry, and computers. This career also requires extensive mapping knowledge and a critical eye for vital field observations. Some field and laboratory assistant jobs are available to applicants with a bachelor's degree.

characteristics: It takes a sense of adventure and willingness to travel to do this job. You have to have scientific discipline, but remember, this isn't just a lab job; you can't mind getting your hands dirty.

Charles "Skip" Cunningham *is an economic geologist for the U.S. Geological Survey.*

How did you get the job?

As an undergraduate student, Charles Cunningham worked as a field assistant on several research projects for the U.S. Geological Survey. After getting a master's and a Ph.D. in geology, he taught for a year and then moved to the USGS to pursue research projects he'd begun while in school.

> **GET SUMMER FIELDWORK TO GAIN EXPERIENCE. STAY ON THE ACADEMIC TRACK, AND TAKE A GEOLOGY FIELD COURSE.**

What do you do all day?

Charles is usually juggling several projects at once, sometimes in opposite parts of the world. He spends a good deal of time traveling to some of the world's most remote spots. It's one of the things he loves about the job: getting to see places tourists rarely get close to.

When Charles travels to a site, his equipment includes maps, geological aerial photographs, a hammer, a hand lens, and a notebook in which to record his observations. At the scene, he's able to make some evaluations instantly from observation and experience. He also collects samples to bring back to the laboratory for in-depth analysis. If you see him in an airport, he's the guy with the heaviest luggage. On the return trip, Charles has been known to bring back a truck filled with rocks!

Back in the lab, Charles studies his samples using some of the most sophisticated instruments available. One recent project, involving mineral deposits in a large volcanic field in Utah, may take years to evaluate. Other projects can be completed in just a few months. One of the jobs involves samples from Colorado, where he's developing new ways to discover hidden resources of gold, silver, and molybdenum. In the office, he synthesizes and compiles all of the information into detailed reports.

Where do you see this job leading you?

Working at the U.S. Geological Survey has given Charles the chance both to try administrative/management work and to jump back into the scientific research route. He says he loves the travel, the fieldwork, and being able to go to work in blue jeans and a flannel shirt. He says the best part of being an economic geologist is getting paid to practice your hobby. He probably won't be changing jobs anytime soon.

40. Geologist—Marine Studies

description: By definition, geologists are experts in the scientific study of the nature, formation, origin, and development of the earth. Marine geologists specialize in studying the 70 percent of the earth's crust that is underwater. One marine geologist tells us he considers his job the study of rocks where "the water is just in the way." Marine geologists are true explorers, gathering and interpreting data to expand our knowledge of the mostly unknown ocean waters. Their job takes them from the office to the laboratory to the great outdoors as they explore all the geological factors of the seafloor. They examine not only terrain but also chemical elements, erosion, materials, processes, and underwater volcanoes and earthquakes.

salary: With a bachelor's degree, the starting salary is about $25,000, with a master's it's about $28,000, and with a Ph.D. it is about $40,000 or more a year.

prospects: As one geologist tells us, "The ocean is a vast area of the unknown; therefore it is a growth field for future research." So where are the jobs for this field?

Jobs can be found in research and teaching for both laboratories and academic institutions. Many of these research institutions are affiliated with universities or with marine study institutions, such as aquariums. However, according to the Geological Society of America, "Academic employment opportunities in the geological sciences are modest but stable." There are other opportunities as well; the oil and gas industry is a major employer of geoscientists. There is a variety of specialties within the field, not only in exploration but also in development of new technology, and environmental cleanup is an area that continues to generate new opportunity.

qualifications: To supervise research projects and get the top jobs, you'll need a Ph.D. However, even with a Ph.D., it is a very competitive field. With only a bachelor's degree, expect to work as a technician on research projects. Technicians tend to do more of the tedious tasks—the grunt work—but the job can still be rewarding.

characteristics: This line of work demands someone with an inquisitive and analytical mind and the knowledge and ability to pose the right questions. Geologists also need to know how to write well, as they're required to communicate and publish their findings. A geologist needs to be self-motivated, diligent, thorough in research, and detail-oriented.

Dan Orange *is an assistant scientist and robotics specialist with a marine research institute.*

How did you get the job?

As a graduate student, Dan Orange participated in several field projects, one of which included working with a submersible (a mini-submarine). That began his journey into the world of underwater geology. He saw his current job advertised in a geologists' trade association newsletter.

What do you do all day?

Dan explores the ever-changing and fascinating ocean floor. He observes undersea places and things never before seen by any human being. The absolute high point of his job is to control a robotic submarine equipped with remotely operated cameras. The submarine is also outfitted with computer sensors that collect data. The "Remotely Operated Vehicle," or ROV as it's called, acts as Dan's eyes, helping him "see" an otherwise unreachable part of the underwater world.

However, before sending down the ROV, Dan spends

> **GET INVOLVED WITH SEA RESEARCH PROJECTS WHILE IN COLLEGE. VOLUNTEER TO GO OUT ON A RESEARCH CRUISE; THEY CAN ALWAYS USE AN EXTRA PERSON. THE JOB MAY BE AS TEDIOUS AS LOGGING INSTRUMENT READINGS EVERY HALF HOUR, BUT THE EXPERIENCE IS CRUCIAL. YOU SHOULDN'T GRADUATE FROM COLLEGE WITHOUT ANY RESEARCH OR PRACTICAL EXPERIENCE.**

enormous amounts of time doing research. He studies maps of the ocean floor that are in part created with sonar devices. These maps indicate underwater fault zones and volcanoes. "This allows me to narrow down an area where something interesting is happening," Dan says. "The ROV allows me to have eyes on the bottom, and with its manipulator arm I can go down and start poking around to see exactly what is causing unusual activity or to see how a fault zone looks."

Dan chooses his own projects, picking ones that spark his interest, but he must also write the proposals and prepare the budgets for each study.

He's not only a scientist; he's also an innovator. His job includes developing the next generation of instruments used for gathering data and for exploration. In fact, his lab looks more like a machine shop than a traditional laboratory.

He does have his share of frustrating days. "Dive days cost $6,000 to $10,000 a day, and sometimes the weather is so bad you don't dive, but the clock is ticking. Or you spend $60,000 building an instrument and it doesn't work, or a cable breaks and you lose it."

Setting up the lab consumed his first months on the job. He was the one to order everything from microscopes to map cabinets to computers. From his previous work, he knew which

projects he wanted to start with. "You may find this hard to believe, but I do get paid to work on what I think is interesting, so rather than being given a project, I will pose some hypotheses based upon the data I have and then propose going to test my hypothesis using the ROV," says Dan. "I knew exactly where my first 10 dives would be."

Now, three years later, he says his work is at a point where he needs to write about his findings and get them published. Published findings, he says, are the way research organizations or institutions are measured and win funding. He also often travels around the country to attend conferences, give speeches, and engage in seminars about his work.

Where do you see this job leading you?

Dan plans on staying where he is, as the ocean research possibilities are endless. He looks out his office window to the sea crashing on the shore and says, "It is a dream job."

41. Environmental Landscape Architect

(RESTORATION)

description: This career is a new offshoot from traditional landscape architecture. An environmental landscape architect designs, plans, modifies, and enhances large tracts of land with an eye toward environmental responsibility and stewardship for the land. Some landscape architects specialize in restoring wetlands and woodlands to their natural habitat; it may be a case of restoring an area around a highway construction or remediating a landfill site. Landscape restorers can turn dumps into parks, and environmentally damaged tracts back to natural marshland. Instead of deciding where a decorative fountain should be placed, a restoration landscape artist concentrates on restoring an entire ecosystem. He or she researches what indigenous plants, animals, and birds are needed to re-create the natural habitat that once survived on a particular site. You've got to know what you're doing, because mistakes can be costly and can destroy the environment you're trying to enhance. For example, in Florida, nonindigenous species, such as Spanish moss, were planted. Today Spanish moss grows wild and is out of control, killing trees and, in some places, entire ecosystems.

salary: Entry-level architects start at $20,000 a year. After several years of experience, top architects can make upward of $80,000–$100,000 a year.

prospects: It's a steadily growing field. The tricky part is finding the firms that offer such specialization. Because it is in demand, prospects are good in this specialization.

qualifications: Obtaining a bachelor's degree in any one of the environmental sciences is recommended, along with a master's in landscape architecture. Summer internships doing wetland delineations, collecting field data for a university professor, and working in a CAD drafting position are also highly recommended. Today's landscape architects must have a strong knowledge of science and be well versed in the concepts and applications of ecology, botany, and biology.

characteristics: This job requires great initiative to learn on one's own. Obviously, you have to love the outdoors and not mind getting dirty. This is a hands-on job.

Bill Young *is a restoration architect based in Florida.*

How did you get the job?

"I created it." Bill Young owns his own landscape restoration company. He started it two years ago and, without any marketing, is doing a booming business. In fact, he just landed Florida's biggest client, Walt Disney World.

Bill's first job was as a landscape architect with New York City's parks department. He helped redesign and construct Brooklyn parks. That led to a transfer to the Department of Sanitation, where his work included supervising the clay capping and covering of a 2,400-acre landfill. On the job he learned all the necessary regulations.

When his wife's job forced the Youngs to move to Florida, Bill decided the time was right to start his own company. Working with a plant nurseryman, a landscape contractor, and an ecologist, his first project was restoring a 13-acre landfill site. The private company that

LEARN ABOUT THE GRASSROOTS BY DOING WORK IN YOUR COMMUNITY. SURELY THERE ARE ILL-DEVELOPED AND ENVIRONMENTALLY DAMAGED SITES IN THE COMMUNITY THAT WARRANT RESTORATION. VOLUNTEER TO HELP. ANOTHER STRATEGY IS TO CONTACT THE SOCIETY OF ECOLOGICAL RESTORATION (SEE RESOURCES, PAGE 217).

owned the landfill had, over the years, dumped dyes and solvents there. Assured that the landfill was more than adequately capped, Bill's team restored the site to a low-maintenance grassland meadow.

What do you do all day?

One advantage of working out of the house is not having to dress up in a suit every day. Bill usually starts his day by making phone calls to prospective clients or researching and collecting data for current projects. He saves research that involves reading for the evenings. Since he has several out-of-state clients, he does have to travel. Bill was recently on cloud nine

because he'd just landed the Disney World contract to restore the canals through which boats transport visitors from one section of the park to another. When Disney World was constructed, much of the wetlands that would have natu-

rally protected the shorelines from erosion were destroyed. "Three days a week I'll be going out into the creek wearing hipwaders and making readings with a measuring rod. Then we'll go to the computer, make a map, and start designing. We are not prima donnas; we get out there with our boots and shovels and we plant." In the meantime he's making calls, finding books, and doing research to find out which plants and grasses are indigenous to the area.

Where do you see this job leading you?

The company has taken off with no advertising and little marketing; it's all been word of mouth. There are countless environmentally damaged sites that need cleanup and restoration. Bill hopes to cash in on that and improve the environment.

42. Environmental Consultant

description: All environmental consultants must have compassion for the environment, but they must also have the ability to interpret legislation and direct their clients in following that legislation. Environmental consultants are usually hired to ensure that environmental regulations are being met during the construction of a roadway, factory, or any other structure that is required to meet these regulations. The consultants are usually brought into a project early on so they can help in obtaining licenses that will allow a company to begin construction. They are also often asked to stay on throughout the project's construction.

Consultants usually work as a team searching out possible environmental problems such as the intrusion of an endangered species habitat or the contamination of wetlands. Once a situation has been identified, they must try to find a solution that will allow a project to continue without disrupting or damaging the environment.

salary: An environmental engineer working for a consulting firm in the private sector can expect to start out making around $30,000. This salary can grow with experience to something in the neighborhood of $60,000 to $70,000. However, if you were to move into a management role, you could make a salary as high as $100,000 a year.

prospects: The future for these types of jobs largely depends on government funding and regulations. Currently, with cutbacks on the horizon, the job market for environmental consulting in the governmental arena looks dim.

qualifications: It is recommended that someone going into this field should hold a bachelor's degree in a broad-based environmental discipline such as environmental studies or environmental engineering. In order to further focus your studies, you should gain a master's degree or doctorate in a specific area such as hazardous materials, endangered species, or water conservation.

characteristics: Whether you are working as a biologist or as an environmental engineer for a consulting firm, you will more than likely have two things in common with your co-workers—a love for the outdoors and a concern for the future of the earth's natural resources.

Paul Sugnet *is an environmental consultant specializing in wetlands ecology and restoration.*

- -

How did you get the job?

Paul Sugnet started his career in environmental consulting by working summers for the Bureau of Land Management while he was earning his bachelor's degree in the conservation of natural resources. He also spent a summer working for a private firm before he began volunteering with the National Park Service and working toward his master's degree. Paul was eventually hired by the Park Service, where he helped conduct a fire-history study of the park. However, after he earned his master's, the market for government jobs was slim, so Paul decided to expand his job search to the private sector. He began working for a consulting firm on the West Coast, where he

gained experience working with wetlands and endangered species. Eventually he decided to do freelance work. Before he knew it, his freelance jobs grew into a business of his own. Now Paul is the president of Sugnet and Associates Environmental Consultants.

What do you do all day?

As the president of an environmental consulting company, Paul spends much of his time in meetings, cultivating new

clients and keeping up relations with current ones. However, he makes it a point to get out in the field occasionally so as not to become too bogged down with the gritty details of running a business. Paul says the hands-on work in the field is where he gets the greatest rewards from his work. "I think the biggest payoff for me is to feel like we are solving problems, that we are beneficial to society.

We do a lot of work that is just red tape, but what we feel good about is when I can say that we accomplished something that is relevant to other people in society," Paul says. "It's when we can measure water quality or when we can say we have 1,000 more ducks or geese coming onto a site than we did the year before, that is when the work is most satisfying."

Where do you see this job leading you?

Paul plans for his company to continue to prosper in the future. He also intends to pursue other ventures related to the environment, such as mitigation banking.

> **ASIDE FROM SCOURING THE NEWSPAPERS TO FIND JOB OPENINGS IN ENVIRONMENTAL CONSULTING, CHECK WITH PROFESSIONAL ORGANIZATIONS SUCH AS THE NATIONAL ASSOCIATION OF ENVIRONMENTAL PROFESSIONALS. (FOR ADDITIONAL INFORMATION, SEE RESOURCES, PAGE 216.)**

43. Wetlands Ecologist

description: Our nation's wetlands have faced the threat of development for many years. But now there is a more concentrated effort to save, restore, and protect these precious parcels of land. A wetlands ecologist studies the biological and physical features of these threatened lands. That includes flood control, wildlife habitat and water pollution control, and other wetlands concerns.

Wetlands ecologist jobs can be found both in the public and private sectors. In the public sector, some ecologists are in charge of managing or monitoring wetlands properties, while others enforce regulations. The job might also involve developing regulations for the management of wetland wildlife species, waterfowl, birds, or endangered species. It may also include reviewing development proposals and applications. In the private sector, wetlands ecologists are primarily employed by environmental consulting companies and large engineering firms. Wetlands ecologists assist in development projects involving wetlands sites, and in creating and designing ways to minimize adverse impact on particular sites.

salary: Entry-level salaries vary regionally, from the low $20,000s to the mid $30,000s.

prospects: Many local and state governments have joined forces with federal agencies to impose stringent wetlands protection laws. Those government regulations and heightened awareness of wetlands importance have opened career opportunities in this field.

qualifications: Training in wetlands ecology or wetlands science can be found in several academic departments, including environmental science, fish and wildlife, geology, biology, and botany, as well as forestry. It is a multidisciplinary field calling for a combined background in many of these sciences.

characteristics: Wetlands science is for someone who enjoys integrating many fields of study like animal habitats, water quality, and hydrology. A wetlands ecologist almost always works as part of a team of engineers, wildlife biologists, and soil scientists, and needs to be able to easily interact with these experts. Good communication and writing skills are essential.

Mickey Marcus *is a senior scientist and partner in a private environmental consulting firm specializing in wetlands.*

How did you get the job?

Mickey Marcus graduated college with a degree in wildlife, but, during graduate school, got a job doing wetlands surveys. He says that that experience was invaluable because, upon graduation, he was easily able to land a job with a private consulting firm.

What do you do all day?

Mickey says his job is a combination of many sciences: botany, geology, hydrology, wetland hydrology, wildlife biology, and other sciences. Although he works with experts in each of these fields, he must have a strong knowledge of all of them.

Mickey usually starts a project by doing a survey of the

> ONE WETLANDS SPECIALIST SUGGESTS GETTING A MORE SPECIALIZED DEGREE, RATHER THAN ONE IN ENVIRONMENTAL SCIENCE. THIS PRIVATE CONSULTANT SAYS THAT AN ENVIRONMENTAL SCIENCE DEGREE IS TOO BROAD, AND THAT EMPLOYERS ARE SEEKING THOSE WITH SPECIFIC SKILLS AND CERTIFICATION.

area. This involves walking through the property, identifying wetlands and resources, and classifying plants and soils. Using soil kits, he'll extract samples, take plant counts, and, if necessary, assess wildlife.

When a property owner, for example, wants to develop 20 acres of land for a parking lot, subdivision, or shopping mall,

Mickey's team may be hired for the initial assessment. The project's designers, engineers, and architects then base their work on the team's surveys and assessments.

"The advice we give people is to try to avoid filling, altering, or degrading wetlands or sensitive habitat. In cases where avoidance is not possible, we try to minimize the alteration or fill with engineering structures, retaining walls, or other methods. The next step would be some sort of mitigation . . . if

you fill 5,000 square feet, then the builders are required to build 5,000 square feet of wetlands elsewhere." Mickey's consulting firm also helps the client through the regulatory permit process.

What Mickey likes best about the job is the variety. Every project is different, offering new challenges. The downside of the job, he says, is working under constant deadline pressure to produce documents on time. "When you're working on several projects, that can be difficult . . . it can be a pressure cooker."

Where do you see this job leading you?

As a principal in the company, Mickey is very happy being where he is.

44. Hazardous-Waste Test Engineer

(PRIVATE CONSULTING FIRM)

description: When hazardous waste has been improperly disposed of or when there is some question about the safety of a particular site, a hazardous-waste test engineer is called in to verify the substances present in a local environment. In this job, you're collecting and analyzing soil and water samples and providing information to people who need to know if their health or the public's safety is in danger.

salary: Expect a salary of $25,000 to $30,000 to start, although an engineering or advanced degree will likely command more money.

prospects: There is a real demand for engineers in this field. The information hazardous-waste engineers collect can determine if proposed construction projects can go ahead and is often used to determine how federal Superfund cleanup money will be spent. The recent closing and realignment of military bases in the United States also represents a boon to private consulting firms that do this type of testing; as bases are decommissioned, they must be deemed safe for new private sector uses. Operational changes at NASA's various research and launch sites also present opportunities for people working in this field.

qualifications: Minimally, a bachelor's degree in environmental science is required, but an engineering degree is preferred. You should have some experience collecting soil and water samples, and a good working knowledge of chemistry and biology. You should be in good physical health, since your lung capacity will be repeatedly tested to ensure that you are able to use an air-purifying respirator.

characteristics: This is not a job for the timid, since some sampling jobs may put you in close proximity with toxic and potentially lethal chemicals. (Of course, you wear full protective clothing.) You must be self-motivated and adaptable. The work can be tedious, but you are likely to encounter a variety of different working environments. For consulting work, it helps to be a quick study: someone who can learn a lot about a client's business or needs on short notice.

Robert Williams *is a senior project engineer with a private waste management consulting firm.*

How did you get the job?

Robert Williams has a bachelor's degree in chemical engineering. He was well on his way to earning his master's in environmental engineering when he began to get reports from former classmates who had moved out into the workforce. They felt unfulfilled working for chemical companies; some said they were "trapped" working in factories or chemical plants. To avoid their fate, Robert pursued a job with a small environmental consulting firm after graduation. He moved to larger firms, and arrived at his current company only after a boss with whom he had a particularly good rapport made the jump and brought Robert along.

What do you do all day?

Work for a consulting firm requires a lot of salesmanship and proposal writing. One day recently, Robert had been filling out a stack of government forms that would enable his company to bid on a series of municipal cleanup projects. At the same time, he was answering questions from a younger staff person who was working on a proposal for the Army Corps of Engineers.

The day before, Robert had been at a potential cleanup site near his home office evaluating the types of samples that needed to be taken. A developer planned to build a hotel near an old hazardous waste site and needed to know what chemicals were present in the surrounding soil and in what quantities. For his most recent on-site visit, Robert wore a hard hat and steel-toed boots, but other sites require the full protective regalia and decontamination procedures designed to protect his health. Most days, though, he's in a business suit, walking the site with a client or regulator.

The extent of sampling also depends on the project. "A lot of sampling involves just dipping a jar into a stream or surface water, and all that's needed is a pair of gloves and a jar. Other times it's just scooping a bit of soil, while other times we go out with drill rigs and drill down into the soil surface."

Robert and his field employees undergo annual physicals and blood tests to make sure that the chemicals they've been around haven't compromised their health.

Where do you see this job leading you?

If he stays with his current employer, Robert sees two career tracks: technical and managerial. He appears to be on the managerial track now, accepting more supervisory and administrative duties as he rises in the organization. However, Robert has begun to think about the possibility of going to law school. His dealings with various projects have underlined what he believes is a serious need in the area of environmental law. For the time being, though, Robert says, he's staying put.

> **IF YOU ARE FOLLOWING THE ENGINEERING CAREER TRACK, TRY TO GET CERTIFICATION. THERE ARE DIFFERENT TYPES OF CERTIFICATION AVAILABLE, DEPENDING ON YOUR EDUCATIONAL AND PROFESSIONAL EXPERIENCE.**

45. Radiological Remediation Expert

(RECOVERY/CLEANUP TECHNICIAN)

description: These are the men and women who detect and clean up contaminated radioactive sites. Contamination is found in countless industrial sites, from facilities that build spaceships to those that manufacture smoke detectors. These recovery technicians determine if there is any contamination, what the contaminants are, and how much there is.

The start of each new job is a challenge. The technician will first research any available information about the history of the site by checking records and conducting interviews. Then he or she extracts samples, offers assessments, and recommends the proper method of cleanup. Wearing protective gear, radiological remediation experts conduct tests and samplings with specialized detection instruments. They usually supervise the laborers who bulldoze the site and who remove contaminated soil or debris. On soil remediation sites, the dirt is removed and packaged for shipment to a disposal facility. Cleanups can last anywhere from a few weeks to months, depending on the size of the project.

salary: Depending on the project, entry-level wages start anywhere from $6 to $12 an hour and can range up to $25 an hour.

prospects: Job prospects have opened in some areas and tightened up in others. For example, the closing of many military bases has forced cleanup of these large sites, which has been a boon to the remediation business. On the other hand, if the federal government, which funds many cleanup sites, decides to raise the limits of allowed levels of contaminants, then there will be less demand for cleanup.

qualifications: Two-year college programs can be enough to get a job as a technician. Science courses and classes in nuclear energy are suggested.

characteristics: It takes teamwork to get these jobs done, so you have to be able to work well with others. You also have to be flexible, as each project presents its own challenges. Often, the job requires travel and being away from your family for weeks at a time.

John Hager *is a radiological remediation expert and recovery-and-cleanup technician.*

How did you get the job?

John Hager got his training in the Navy. After leaving the service, he got a job as a radiation protection technician in nuclear power plants. His job was to monitor radiation contamination around the plant, making sure there was none.

John also monitored employees who worked on radioactive and survey equipment. As a temporary employee, he traveled around the country, working at many different plants. There, John met employees of the company for which he now works, and through word of mouth heard about the job he currently holds.

What do you do all day?

Recently, John was in the office catching up on paperwork, but usually he's in the field on site. He spends most of his time away from home and out on cleanup project sites.

His team begins each assessment by talking with property owners and sifting through paperwork to find out the history of the site: operations that occurred there, what chemicals were used, and how much. Depending on the levels of suspected contamination, John wears protective clothing. "Sometimes you have to wear respirators, sometimes you have to dress in full anti-contamination clothing. You don't have to do that often, because most of the stuff is fairly low level. On most jobs, just shoe covers and gloves are enough," he says. Using radioactive-detection instrumentation that usually consists of small handheld instruments such as Geiger counters and ion chambers, John tests and collects samples. After an appraisal of the pollutants, his team offers a cleanup plan. Some of the sites are 20 acres and larger.

"It's never routine," John says. One cleanup that stands out in his mind was a Superfund site in Philadelphia where houses had been built with radium-filled concrete-sand mixtures. "In some houses it was just one wall where the radioactive mix was used … other houses had to be totally gutted." Does his work ever scare him? "No, I'm not afraid of it. I respect it. We have instruments that tell us it's contaminated before we even get to it. You should never walk into an area and not know what's there," he says.

> ONE SUGGESTION FOR ENTERING THIS FIELD IS TO TRY TO GET A JOB AT A NUCLEAR POWER PLANT. RADIATION TECHNICIANS ARE OFTEN TRAINED ON THE JOB, AND SUCH A POSITION MAY BE YOUR ENTRY INTO THE RADIOLOGICAL REMEDIATION INDUSTRY.

Where do you see this job leading you?

John isn't looking to move to another job. He likes what he does and is anxious to move on to the next project.

46. Biotech Remediation Project Scientist

description: Nobody ever intends for a petroleum leak or spill to take place, but when one does occur, there has to be someone who will plan the cleanup so that the spill will not destroy other natural resources such as drinking water or soil. That is what a project scientist working in biotech remediation does. When there is a spill, he or she immediately goes out to it and begins assessing the damage. This involves taking samples from both soil and water. Once the amount of damage caused by the spill is determined, a written report must be prepared and sent to the state or other regulatory agency, detailing the damage and the plan for cleanup.

salary: Someone starting out as a field technician, which is a typical starting position for the industry, will be making in the range of $18,000 to $22,000 a year. After gaining experience, a field technician will typically move on to the position of a project scientist and bring in a salary of $24,000 to $30,000 a year. A senior scientist, the next step, can earn as much as $30,000 to $45,000 a year.

prospects: Any time the state or federal government creates a regulation, someone must ensure that people stay in compliance. With the number of environmental regulations increasing, there will continue to be jobs for people who monitor compliance.

qualifications: Earning a bachelor's degree in the geological sciences or hydrogeological sciences would be a great start for anyone wanting to break into this field. But it is important that while concentrating your studies for your degree, you take a broad range of course work in the sciences so you will have a working knowledge of many areas. This course work must be coupled with some type of experience in the field, which can be gained through a work-study program, a summer job, or an internship.

characteristics: It is important to be able to work well with people in this job; you must be able to keep a good working relationship with the clients for whom you are conducting a cleanup, as well as with your colleagues. It is important that you be able to turn to your colleagues for assistance in their fields of expertise. At the same time, you must be able to offer your expertise in a nonthreatening manner.

Ross Kennemer *is a biotech remediation scientist who works for a petroleum company.*

How did you get the job?

Throughout college Ross Kennemer took various courses in the sciences because they interested him, but he admits that he didn't have a specific goal in mind as far as what type of career this course work would lead him to. "I took geologic science courses, a fair amount of chemistry, and physics. I took math through calculus and I took a lot of soils courses. I really diversified," Ross remembers. "I just took what courses interested me. I did this unknowingly. It was just to satisfy my curiosity and my interests. I had no idea what I would be doing today."

After graduation from college, Ross decided to follow his interests again by looking for work in the petroleum industry. He very quickly found work at a gas-producing company in western Texas. From there, contacts that he made led him to a job at a soils engineering firm,

THE BEST WAY TO FIND A STARTING PLACE IN THIS FIELD IS TO LOOK THROUGH YOUR LOCAL PHONE BOOK FOR ENGINEERING FIRMS. CALL THE FIRMS AND SEE IF THEY NEED INTERNS OR SUMMER WORKERS. ANOTHER GOOD IDEA IS TO LOOK IN ENVIRONMENTAL REGISTRY, WHICH CAN BE FOUND IN ALMOST ANY LIBRARY. THIS RESOURCE BOOK WILL PROVIDE YOU WITH A LISTING OF ENVIRONMENTAL FIRMS.

then a job at a bioremediation firm, before bringing him to the point where he is now.

"I haven't ever gotten a job out of a newspaper or out of a want ad or anything like that. My career path has just flowed on its own. To me it has been a natural progression. In each place I have progressed up, I have progressed up to a point where there was really no other place to go."

What do you do all day?

As a project scientist, Ross spends about three days at a time out in the field working on a spill. But for every three days in the field, he can expect to spend about a week or two in the office writing a report to be submitted to the state or other regulatory entity. Typing up reports can be tedious, and while it isn't what Ross enjoys most about his job, it is an essential part. Typically, there are more jobs in the spring and summer, which means these are the months that Ross will be working the longest hours. In the summer, it is not rare for Ross to work a 15-hour day in order to meet a deadline for a particular job. Things will slow down in the wintertime, but won't stop completely.

While out in the field, Ross manages the employees who are drilling and collecting soil samples. This type of management requires that he make numerous

decisions about the progression of the cleanup throughout the day. These decisions must be made quickly so that the deadlines for each job can be met. "I am under a time line on everything. I have a set time that a job has to be performed and a report has to be prepared and submitted." But, Ross says, the pressure of a deadline doesn't necessarily create a stressful situation. "I enjoy the pace that I go at; I don't consider it to be highly stressful."

Where do you see this job leading you?

Ross plans to progress in his field, but he is not restricting himself to the bioremediation cleanup of underground storage tanks. "Environmental science is my passion that I have developed, and this is where I see myself working, but it is so diverse and broad. Next year it may not be underground storage tanks, it might be air emissions or it might be strictly groundwater."

47. Field Sampling Supervisor

(PRIVATE LABORATORY)

description: A field sampling supervisor oversees the taking and preservation of soil, water, and material samples that are to be tested for chemical content. In this job, you are a player in the "compliance game": the need for potential polluters and permit holders to make sure they're staying within federal, state, and local pollution guidelines, as well as the need for environmental law enforcement agencies to make cases against suspected violators. In both instances, samples need to be carefully collected and analyzed according to strict legal protocol, since the information may be used to grant or deny permits, or be introduced as evidence in a court of law.

salary: Depending on the geographic area and responsibilities, entry-level salaries can range from $18,000 to $30,000. Candidates with exceptional potential may be hired at above entry-level. Most remain in the initial salary range for two to three years before advancing.

prospects: Good. Environmental consulting, engineering, and testing laboratories all hire people for this type of work. Their clients include local governments, private industry, and developers, as well as law firms. Radon testing has become especially important for the real estate industry in many areas.

qualifications: Minimally, a bachelor's degree in environmental science or a related discipline such as chemistry or biology is needed. Experience gathering samples is a big plus. You must be in good health, as annual health testing is mandatory.

characteristics: Flexibility is the key. Work hours and amount of travel will vary. Samples, for instance, may require a storm event, which is difficult to predict. It helps to be physically strong.

Donald Cochrane *is a field sampling supervisor for a private environmental laboratory.*

How did you get the job?

Donald Cochrane was educated to be a fisheries biologist. He has a bachelor's degree in marine-freshwater biology. He says he found the job he has now through "networking," which means he noticed friends and associates moving into this line of work and decided to join them.

What do you do all day?

Travel. On the day of this interview, he had already traveled 200 miles visiting five collection sites, including a wastewater-treatment plant, a government office that makes electronic guidance systems for NASA, and a printing plant. Recently, he had to test at a hospital that had flushed large concentrations of disinfectants and other cleaning solutions while washing down a room that had been occupied by tubercular patients. He often wears protective gear: anything from safety glasses and splash bibs, to full protective "moon" suits with respirators. Samples have to be carefully gathered, marked, and preserved until they are in the hands of the laboratory. Donald says in all cases he must be aware that he is establishing a "chain of custody" for the sample that must stand up under

GOING INTO THE INTERVIEW, HAVE A CLEAR IDEA OF THE JOB YOU WANT. DON'T TELL THE INTERVIEWER, "I'LL DO ANYTHING." INSTEAD, BE CONFIDENT AND FOCUSED. BUT BE FLEXIBLE AND WILLING TO CONSIDER A SIMILAR JOB OPPORTUNITY WITHIN A COMPANY.

scrutiny. Much of the sampling is unannounced, according to federal EPA protocol. Test vessels can range in size from test tubes to large drums.

Where do you see this job leading you?

Donald is already moving along the career track in his company. He would like to begin work on an advanced degree and attain a managerial position with greater authority. He believes he is able to supervise support, clerical, and marketing people as well as other environmental scientists.

48. Pollution Enforcement/ Field Sampling

(ENVIRONMENTAL PROTECTION AGENCY)

description: Field samplers are the pollution police. Enforcing federal government pollution laws begins with literally testing the waters. Field samplers check the discharge from companies that are known to use large quantities of possible contaminants to manufacture their products. People in this job may also follow up on complaints or other inquiries. While some samplers investigate specific companies, others monitor oceans, rivers, lakes, and streams for pollution. EPA testers show up unannounced at companies. Once they explain their purpose to the company chief, the work begins. They check company records, permits, and orders, and then collect samples, conduct and evaluate analyses, and write reports. These reports are later reviewed to determine what action, if any, should be taken, and whether citations or fines should be levied. Some cases are taken to court.

salary: Starting salaries for this work are around $20,000.

prospects: Unfortunately, jobs with the EPA follow the ebb and flow of the political climate. Since the EPA has recently been threatened with budget cutbacks, the outlook for increasing the ranks is not good.

qualifications: A strong science background in such areas as soil science, water quality, biology, natural resources, or environmental science is recommended. Because the work involves evaluating industrial processes, an engineering background is also helpful.

characteristics: This job demands creativity because each inspection presents different challenges. While sampling procedures follow certain protocol, the tester often has to devise ways to extract the samples. Because samplers may be called to testify before hearings, they must also be good speakers. Since companies aren't always happy to see the EPA at their doorsteps, cooperation isn't always forthright. To get the job done, you've got to be able to get people to cooperate. You've also got to be in good physical shape; it takes a lot of physical strength to carry all of the sampling equipment.

Michael Glogower *is an EPA field sampler/enforcement officer.*

How did you get the job?

Michael Glogower didn't start out wanting to do this work. He graduated with a degree in natural resources and applied for a job with the government. He ended up getting hired by the EPA as a soil scientist doing environmental impact studies on sewage treatment plants. Working for the EPA created several challenging opportunities, including policing ocean dumping. Eventually, he was transferred to his current position as a field sampler/enforcement officer.

> WHILE THE PROSPECTS FOR OBTAINING FEDERAL GOVERNMENT JOBS AREN'T GREAT, DON'T LET THAT DISCOURAGE YOU. YOU MIGHT HAVE BETTER LUCK GETTING REGULATORY EXPERIENCE WITH CITY, COUNTY, OR STATE DEPARTMENTS OF ENVIRONMENTAL PROTECTION.

What do you do all day?

A large amount of preparation is needed before an enforcement officer heads to a site. Michael has to review files, get a permit, and learn the history of the facility being investigated. Sampling is done in accordance with the permit.

Michael then determines how many assistants he'll need and what equipment to take. For some jobs, automatic samplers equipped with timing devices are used; other samples are taken manually. The sampling is done following specific procedures and guidelines, and the inspector usually wears protective gloves, a hard hat, and safety shoes. Michael did have an inspection that required him to sample in a wet well, which was about 20 feet deep, and use a self-contained breathing apparatus.

Typically, it takes two days to inspect each facility, but that varies with each location.

Once the visit is completed and samples are analyzed, Michael writes an extensive report, including any recommendations or steps necessary to correct a pollution problem.

Where do you see this job leading you?

Michael is sticking with the EPA. He says the best part of the job is getting out and learning firsthand how industries operate. But his greatest satisfaction comes from knowing that he is cleaning up and protecting the environment.

49. Sanitation Police Officer

(CITY GOVERNMENT)

description: A sanitation police officer is a cop with a specialty. In many jurisdictions, the job of enforcing health codes and sanitation laws falls to people who write summons but lack arrest powers. In large cities, their jobs may be augmented by specially trained police. These armed officers investigate cases of illegal dumping, improper disposal of toxic wastes, and even unauthorized disposal of tires and derelict vehicles.

salary: In many cities and communities, sanitation law enforcement officers start as sanitation workers— working regular garbage pickup. Those starting salaries are around $20,000. New York City sanitation officers are some of the highest paid in the country. They start at $30,000, plus overtime.

prospects: The number of these positions depends on local political priorities. High-profile incidents such as medical waste washing up on public beaches usually result in additional money for this type of work. Officers report heavy caseloads and not enough manpower.

qualifications: Law enforcement training and a criminal background check are required. You must demonstrate a proficiency in weapons use and be physically fit. An understanding of public health and sanitation issues is a plus.

characteristics: This is not a job for the timid. You need to be alert, resilient, and open to flexible hours. You will be working with the public and interacting with other police officers in what are often less than ideal working conditions. There is some desk work involved.

Efraim Acosta *is a sanitation police officer with the New York City Sanitation Department.*

How did you get the job?

For 10 years Efraim Acosta was a New York City sanitation worker. He drove a garbage truck most of the time and in winter months operated a snowplow. He saw an opening posted for his current job and applied. It took a civil service test, six weeks of police academy training, and a physical exam to get in.

> **CONSIDER ENROLLING IN CRIMINAL JUSTICE COURSES.**

What do you do all day?

What he does all night is the more interesting question. Efraim works the night patrol, a steady 3-to-11 shift that might require additional overtime if he makes an arrest. He started his career with the Sanitation Police on the Derelict Vehicles Task Force, tracking down abandoned vehicle owners. But now he specializes in catching illegal dumpers. He and his partner sit on stakeouts in the city's back alleys and lots, and the work doesn't end with the arrests. A day may involve convincing an assistant district attorney to file harsher charges against an illegal dumper.

There's no question the job can be dangerous. Recently, he says, he chased an armed suspect into an abandoned building. Another time, a man was stopped from dumping what was thought to be a barrel of toxic chemicals; it turned out to be a dead body! "We have a front-row seat to the world," Efraim boasts. "Every day is different and exciting. Some people come on board thinking they'll just write health code violations and cite people for not having their dogs on a leash, but it's not that easy. The dog might be a pit bull."

Where do you see this job leading you?

Efraim loves his work. He plans to retire in five years, but says the career path in his department could lead others to supervisory positions.

50. Recycling Coordinator

description: In places where recycling is not mandatory, local governments or nonprofit groups sometimes hire a recycling coordinator to encourage voluntary participation in appropriate recycling efforts. A voluntary recycling coordinator, in some ways, seeks to convert, cajole, and commiserate in order to get people to "see the light." The coordinator's job is also part research—staying abreast of developments in the field—and part adviser, helping the local government or nonprofit groups draft policies and legislation to encourage citizens and businesses to recycle.

salary: Salaries range from $25,000 to $35,000 depending on the degree of experience and the geographic location.

prospects: Interest in recycling seems to have peaked around 1993. Some cities are actually cutting back their efforts. There are now more jobs available with private garbage collection and recycling companies as well as with large firms that want to implement their own recycling programs. A few years back, competition was fairly light, but now applicants for open positions are displaying increasingly high experience levels for jobs that require very little experience.

qualifications: Schools do not offer many formal recycling-education classes, so many companies look for educated people with transferable backgrounds such as business, communication, and marketing. Although knowledge of the recycling field is required, it is something that can be researched outside the academic arena. Environmental or science degrees are not mandatory. It does help to have some experience with recycling at the business or volunteer level.

characteristics: Management and communication skills, along with an analytical mind, are key to making this job work. As a recycling coordinator, you'll be creating ways to convince people to recycle. It's a matter of staying in touch with developments in the field and devising ways to apply various methods and programs to the people and businesses in your geographic area. This is a job for self-starters.

Terry Engle *is a municipal coordinator for commercial recycling.*

- -

How did you get the job?

Terry Engle has a degree in marine environmental studies, but was never able to find a job that suited him in that field. He became interested in the business end of recycling during his 15 years in restaurant operations and management.

"The issues in the field are not relatively complicated," Terry says. "You don't have to be a rocket scientist to figure them out. If you study them for a while, ask the right questions, and if you've got enough on the ball, you'll figure out what has to be done."

Terry says he first identified organizations that dealt with recycling issues, and followed up with research at the career center. Informational interviews and lots of networking helped him hook up with trade associations and other industry contacts. He volunteered at a recycling center, where he saw his current job announced.

What do you do all day?

Much of it is like being a project manager. Terry supervises three employees. He reviews contracts and manages outreach projects and research efforts. He's always looking at ways of improving existing policies or motivating residents or businesses to recycle more.

Terry's first days on the job were consumed with switching contractors for the collection of recyclable paper in municipal offices in San Francisco. The new contractor was apparently also new to that type of work, and Terry had to keep him on track. "It was trial by fire," Terry says.

Terry spends a good amount of time with the city's public relations people, brainstorming new, creative ways to get out the message about recycling. This is a challenge, since they don't have a lot of money to spend on radio or television. Instead, Terry says, they concentrate on outreach efforts through various organizations or associations.

> **GET INVOLVED IN RECYCLING EFFORTS IN YOUR COMMUNITY. YOU'LL LEARN ABOUT THIS NEW INDUSTRY AND ALSO MAKE CONTACTS.**

Some days Terry may be called to testify before the county commissioners or planning board regarding a recycling ordinance he's researched and drafted. He just finished drafting an ordinance that requires new buildings to include adequate space for storing recyclables, and recently was rewriting an ordinance requiring city departments to purchase recycled material.

Where do you see this job leading you?

There are opportunities both in government and in the private sector, working for a trash collection or receiving company, for a large firm with recycling programs of its own, or as a consultant.

51. Recyclable Material Collector/Broker

description: In some places where recycling programs are not required or operated by the local government, private nonprofit groups have stepped into the breach. The people who run these groups arrange for the collection and sale of recyclable material, using the profit to pay their operating expenses. In this job, you are convincing people to separate their recyclables and to allow you to pick it up. Then you sell the material to recyclable products manufacturers.

salary: Private sector recycling brokers can start out earning $20,000 to $25,000 annually.

prospects: Good. Even where recycling laws have not been enacted, local governments are under intense pressure to reduce the flow of trash into landfills. In an era when minimal government is considered a virtue, private efforts can flourish. Recycling is a fairly new business, so you must keep up with changing laws and regulations.

qualifications: Some management experience and a knowledge of how businesses work is extremely helpful. You will need good communications skills and a working knowledge of the recycling field. Recyclers continuously seek out new markets and buyers. You're competing for pricing and supply. A background in business/marketing is a big plus. At the entry-level, a strong back doesn't hurt either.

characteristics: In this line of work you have to be committed and persuasive, yet measured and methodical in how you convince people to recycle. Many people are still resistant to the idea. You must be able to look at the problems of an administrator or businessperson and rationally work out a way he or she can participate in your effort. Beyond that, the nuts and bolts of making the pickups and deliveries is like running a business.

Shari Stern *is an intern at a private nonprofit recycling collection group.*

How did you get the job?

After graduating from Rutgers with a bachelor's degree in human ecology, Shari Stern was headed to graduate school at Tufts University in Boston. She hit the library, got a Boston-area phone book, and combed both for environmental jobs in the recycling field. Every company that seemed to fit the bill received one of her resumes in the mail. Once in Boston, she "crashed" an environmental expo because she didn't have the registration money. She set up an interview with Earthworm, one of the companies she had read about, and got the job the next day.

What do you do all day?

Shari says she's the only employee at Earthworm who doesn't have a trucking license, so she does everything but drive. That includes loading paper from the 870 companies Earthworm collects from and dropping off the material at the

> **START SUBSCRIBING TO TRADE JOURNALS OR MAGAZINES RELATED TO THE SPECIALTY YOU WANT TO PURSUE. THESE WILL ADVERTISE JOBS, AS WELL AS CONFERENCES AND SEMINARS, WHICH OFTEN HOLD JOB FAIRS.**

commercial bailer. It's heavy work, but it's part of learning the job. The outfit also collects toner cartridges from laser printers and sends them to a company for refilling. On days when they're not collecting, there's plenty of office work to be done. They also arrange for the sale of recycled paper to some of the offices they deal with, and they even go to schools to teach children about recycling and composting.

Recently, Shari was working on a study, compiling information about area recycling programs—specifically about those at area universities. Nearly every day she gets calls from people, usually schools or church groups, who want to know how to set up recycling programs. Sometimes callers just want advice on how to convince their boss to create a recycling pro-

gram. Most of the work Shari does is in the office—answering phones, taking orders for pick-ups, answering questions about recycling and her organization's program.

Shari says the biggest benefit of the job is that she learns about other organizations that do environmental work. "I find out what efforts are being undertaken all over the state. I also learn about the recycling market and the business," she says.

Where do you see this job leading you?

Shari says it's too early to tell where this job'll take her. Currently, it's helping her pay the bills as she works on a master's degree in urban environmental policy, but she loves the fact that she's doing something positive for the environment.

52. Salesperson for Recyclables

description: A relatively new industry involves the sale of products made from recyclable materials. Recycling is becoming part of our everyday lives, as in many localities it is now required by law. The technology to manufacture products from recycled newspapers, plastics, and glass continues to improve and flourish. Established waste companies are expanding their businesses to include the sorting and melting-down capabilities necessary to fabricate new products from old materials. Many established industries, such as paper mills, are branching out to manufacture these new products. Smaller entrepreneurs are opening their own businesses and selling these products. A sales manager for recyclables, as in other industries, identifies a customer market and develops a sales strategy. These sales jobs are found in the retail as well as in the manufacturing end of the industry, such as paper mills and plastics manufacturers. There are also wholesale merchants of these products.

salary: Sales is usually a commission-based job. Although these jobs usually offer a minimum base salary, commissioned earnings are dependent on the economic marketplace, the product, and salesmanship.

prospects: Because many companies are willing to give less experienced people a chance, sales has a lot to offer in terms of job prospects; since the jobs sometimes offer low pay and depend on commissions, small companies don't have as much to lose by giving a less experienced person a shot at the job.

qualifications: Salespeople come from different educational backgrounds. Obviously, a strong business sense is key. So is self-motivation.

characteristics: Perhaps the most important qualities for a salesperson to have are persistence, persuasiveness, and ambition. Tenacity and a sincere interest in the environment are also important.

Linda Dempsey *is a salesperson for a recycled paper company.*

How did you get the job?

Linda Dempsey was working for a waste collection company, heading their recycling division, when her current boss called and offered her a job. That may sound easy, but Linda has more than paid her dues.

Her venture into recycling started quite by accident. Linda first became interested in recycling when friends she lived with created an organized neighborhood recycling effort. They urged neighbors to separate their garbage into paper and plastics, which the group collected and hauled to recycling centers. When she moved to Manhattan, Linda kept up the recycling momentum. Before New York City had mandatory recycling laws she urged her neighbors to recycle. Linda herself hauled the paper and plastics to a recycling center several blocks away. She met

> **CONVINCE A PROSPECTIVE EMPLOYER TO GIVE YOU A SHOT BY OFFERING TO WORK FOR COMMISSION ONLY ON A TRIAL BASIS.**

with the neighborhood block association, and before she knew it, she was doing recycling full-time.

Once Linda started charging a nominal fee and taking out advertisements, the business quickly grew. Her clients included commercial businesses from whom she collected used copy paper. But business soon got ugly. Linda says she received violent threats from carting companies, who accused her of stealing their business and forced her to dissolve her company.

Luckily, one waste collection company (not one of those that had threatened her) hired her to

create the recycling arm of their business. First, they wanted to know how much of the trash they collected was recyclable. So for weeks Linda rode on the back of garbage trucks all night eyeballing how much of the garbage could be recycled. She also devised recycling systems in buildings. The job was far from glamorous, but she made contacts throughout the recycling business. Just when she was ready to quit, a new job offer popped up. Now, she's made the transition to sales of recycled products.

What do you do all day?

Linda says she's finally found her niche in sales. The small company she works for sells recycled paper and stationery. They also offer printing. At the lower Manhattan storefront office, Linda keeps her desk near the front door so that she's ready to grab walk-in customers. She has developed most of her clients through her own initiative by working the phones and hitting the pavement. She

conducts research, looking for new markets, and follows up on her own ideas for finding new customers. Recently, Linda had decided to target the numerous art galleries near her office and was headed out to show them examples of her company's printing and paper. Linda not only is busy with customers, but also is on the phone with merchants and paper mills, tracking prices, supply, and new products.

Where do you see this job leading you?

The industry is changing every day, and Linda's ready to get the edge on any new products. The latest is what's called "tree-free" papers made from a combination of hemp straw and oats. Another new paper is made from banana fibers.

53. Recycling Products Manager

description: The individual industries for plastics, paper, glass, and aluminum all have systems and processes in place to "melt down" recyclable materials and create new products. Some waste haulers and recycling facilities are also starting to install these mechanical systems, which sort, clean, and break down the discarded materials and turn them into new products. There are jobs for managers to oversee these operations and for engineers to update and create new mechanical systems for handling recyclable materials. In some cases, these jobs are combined into one.

Many cities now have mandatory recycling laws that require residences and commercial businesses to separate plastics, paper, glass, and aluminum from other garbage. Once the recyclables are collected and taken to a central facility, they are then sorted by hand and with mechanical separators. Many recycling centers, in turn, sell these raw materials to brokers, or directly to industry, where the process of turning used products into new ones begins. It is the job of the operations manager to keep records of incoming shipments, supervise staff, and oversee the maintenance and safety of mechanical systems.

salary: Salaries for engineers and managers in this field start anywhere from around $22,000 to $30,000.

prospects: The prospects for this type of work are very good, as more municipalities are passing laws requiring recycling. There are also areas of new technology that await to be conquered. For example, there are certain grades of plastics and other materials that are not easily broken down. Engineers are searching for new processing and manufacturing systems for these types of products.

qualifications: An engineering degree, particularly in mechanical engineering, is recommended if you want to get into the design and operation of recycling processing technology. Environmental science courses are also helpful. You must also stay up-to-date with changing environmental and recycling laws and regulations.

characteristics: The job takes someone who is vigilant in overseeing processes and production, and who is innovative and imaginative enough to design new systems.

Glenn Lostritto *is a recycling products manufacturer in New York City.*

How did you get the job?

After getting degrees in constructional engineering and business management, Glenn Lostritto entered the family-owned waste collection business. The company, which was started by his grandfather, was ready for Glenn's ideas for updating and expansion into the recycling industry: they decided to take a gamble and built a recycling facility, including mechanical separators to sort the plastics, glass, aluminum, and paper. When New York City passed mandatory recycling laws in 1989, Waste Management, Inc., their company, was ready, and won the city's bid as a recycling collection center.

What do you do all day?

On a typical day, Glenn is making sure that the "processes are running correctly," that is, ensuring that all of the recycled materials hauled into the facility are being expeditiously sorted and separated. But more of Glenn's time is filled with creating new systems to take separated recyclables, grind them down, and turn them into new products.

> BEGIN TO LEARN ABOUT THE RECYCLING BUSINESS BY VOLUNTEERING AT A LOCAL FACILITY OR GETTING INVOLVED IN COMMUNITY RECYCLING COLLECTION PROGRAMS. IT'S A GREAT WAY TO BEGIN MEETING CONTACTS.

Glenn tries to take existing technology and put his own spin on it to fit his company's needs. "We get an idea, and may hire someone to do a feasibility study. I'll do the preliminary designs and take those to a company that actually builds the equipment.

"For me, it's an exciting field because you're using your imagination to create. You can create different products and, at the same time, you're doing something good for the environment," Glenn says.

Where do you see this job leading you?

Glenn's constantly searching for new ways to expand the company and to add value to the recyclables they process, and to the raw materials they produce.

54. Laboratory Research Scientist

description: The job of laboratory research scientist immediately prompts a picture of men and women dressed in white lab coats surrounded by beakers and flasks bubbling over the heat from Bunsen burners. The research opportunities are as vast as human nature's sense of desire for discovery. Lab research demands a diversity of disciplines and is the foundation of man's evolution, breakthroughs, and progress. Research scientists are in demand in all aspects of environmental study, from developing microorganisms that can be used in cleaning up oil spills to creating new fuels. It is a world of constant experimentation and testing of hypotheses. Laboratory research scientists work for corporations, hospitals, technology firms, and manufacturers, just to name a few.

salary: Generally, salaries in this field depend on the level of education. A research scientist with a bachelor's degree might start out at about $25,000, but one with a Ph.D. might earn somewhere between $50,000 and $60,000 a year.

prospects: Job opportunities depend on the field of research. For example, the chances of finding work with a private company doing biotechnology research or genetic engineering are better than trying to get work at one of the national labs. National labs' funding tends to be contingent upon the political climate, and they are now experiencing downsizing.

qualifications: You don't need a postgraduate degree to do this work. However, such a degree provides greater opportunity for better jobs and higher salaries.

characteristics: Patience is the quality most common among researchers; experimentation can be tedious and results slow in coming. You must be able to work alone and be disciplined. One research scientist tells us that you also need a sense of humor to deal with the frustrations that often accompany this work. "Not everything goes the way you'd hope. Strange things happen, like the power goes off on your experiment in the middle of the night, or you might work on an experiment till two in the morning and it blows up."

Tammy Kay Hayward *is an associate biotechnologist at a national renewable energy laboratory.*

How did you get the job?

Determination and a little luck is how Tammy Kay Hayward got the job. She saw the position posted on a school bulletin board. Although she didn't meet all the requirements, that didn't stop her from applying. She was just about to graduate with a bachelor's degree in microbiology, but the advertisement said they were looking for someone with a master's degree. She interviewed and was given the job.

What do you do all day?

Tammy Kay is working on creating alternative fuels. She conducts fermentation experiments, researching new methods to produce fuel ethanol, an alternative fuel to be used in many types of engines. "We can make it from different types of trash components like white paper, magazines, cardboard, leaves, wood chips, and more," she says.

On most days you can find Tammy Kay in the laboratory working at tables filled with small beakers and flasks. However, on other days she writes reports or attends conferences and seminars. She is often called upon to give talks about her research.

Tammy Kay recalls that the first few months on the job were extremely tiring. She spent long days and nights collecting data. "In microbiology, the bugs are growing and once you start a culture you have to be there at two in the morning, four in the morning . . . every two hours to collect samples," she says.

She enjoys traveling to seminars, she says, and the prestige that working at a national lab gives her. The young researcher says she finds it especially rewarding when her work is published. Published findings are critical to research scientists' careers; it is how they and their projects are evaluated and sometimes funded.

Where do you see this job leading you?

Tammy Kay says there are several different avenues that she can pursue. The thought of climbing up the management ladder appeals to her. She realizes that she'll eventually have to go back to school for a postgraduate degree if she wants to really move up and be marketable. With her microbiology degree she also has the option of working for one of the many biotechnology companies or biopharmaceutical manufacturers.

> CONSIDER APPLYING FOR A JOB EVEN THOUGH YOU MIGHT NOT MEET ALL THE POSTED REQUIREMENTS. FOR EXAMPLE, EVEN THOUGH THE JOB LISTING STIPULATES THAT A GRADUATE DEGREE IS REQUIRED, AN EMPLOYER MIGHT BE CONVINCED TO HIRE SOMEONE WITH ONLY A BACHELOR'S DEGREE AT A LOWER SALARY.

55. Windfarmer

description: Generating power through wind can be an exhilarating profession. But finding business, if you are breaking into this field on your own, is often the first and hardest step to take. Because the field of windfarming is so new, it is often hard to gain employment working for a windfarming company. Many farmers break in by starting their own business. Because of the entrepreneurial spirit of this fairly new profession, a windfarmer will spend about as much time in business negotiations as he or she will working with wind turbines. It takes a lot of perseverance, but the payoff for some windfarmers is the knowledge that all of their hard work is going toward a clean, unlimited energy source.

salary: Pay in this field starts low. In fact, a windfarmer beginning his or her own business may not bring in a profit until well into the third year. However, if an engineering position can be secured within a larger, already established windfarming company, salaries will be competitive with windfarmers' counterparts in other fields—in the range of $40,000 to $60,000 a year.

prospects: Jobs in windfarming are hard to come by. There are only about five large wind companies in the United States—located mainly in California—and those companies consist of about 300 to 500 employees. There are also about 100 smaller companies, but they consist of only 5 to 10 employees. Most entering the field will start their own businesses.

qualifications: Aspiring windfarmers should hold a bachelor's degree in engineering, specifically aerospace or electrical engineering. Finding an entry-level job that will allow you to get experience with wind turbines, with utility field service, or with oil company fieldwork may be hard, but an entry-level job working as an aerospace or electrical engineer could give you the background needed for building efficient wind turbines.

characteristics: Windfarmers must be flexible in being able to work both outside and indoors. Paperwork can become tiresome, but that is the only way to keep the customers and the money coming in. Windfarmers must also like to travel, and, because wind doesn't always show up in highly populated places, an appreciation of wide-open spaces is necessary.

Walter Hornaday *is a windfarmer based in Austin, Texas.*

How did you get the job?

With a degree in civil engineering, Walter Hornaday started his career by doing construction work on the installation of coal-fired power plants. But shortly after he started that job, he knew it was one that he didn't want to stay in. "What turned me away from it was the reliance on the depleting natural resources and the pollution side of it," Walter says. He returned to school, this time to get his master's degree, and, while there, began exploring alternative energy sources such as wind.

Armed with no more knowledge than what he read in trade magazines, Walter struck out to get some hands-on experience with wind turbines. He began by driving around southern Texas and looking for abandoned wind turbines. "Wind turbines are pretty easy to spot, so I would drive around and if I saw one, I would just pull over

> THE TRICK TO BREAKING INTO THIS FIELD IS TO FIND THE COMPANIES. THE BEST WAY TO DO THIS IS TO CONTACT THE AMERICAN WIND ENERGY ASSOCIATION. THIS ORGANIZATION MAY BE ABLE TO TELL YOU OF A REGIONAL RENEWABLE ENERGY ASSOCIATION NEAR YOU. FROM THERE, YOU CAN FIND OUT ABOUT WINDFARMING COMPANIES IN YOUR AREA AND TRY TO GET SOME EXPERIENCE IN THE FIELD.

and ask the owners what kind of wind turbine it was and if they wanted me to get it running again."

This method of "job hunting" allowed Walter to get experience with wind turbines, and at the same time gave him the opportunity to negotiate with utility companies. "A lot of times there were abandoned wind turbines from years before, and I would get them up and running again. Then I would work with the utility company to get it interconnected," he says, explaining that the connection allows people receiving energy from the turbine to offset their electric bills.

Before long, Walter started buying dysfunctional wind turbines. Once he got them working, he sold the electricity they generated. His business has since grown, and he is now the part owner and president of Texas Wind Power Company, located in Austin, Texas.

What do you do all day?

On a typical day, Walter spends half of his day in the field and the other half in the office. The morning is usually spent indoors making phone calls to utility companies as well as to potential new customers. He also spends part of this time ordering parts and organizing out-of-town jobs. But the fun stuff doesn't come about until late afternoon, when he goes out into the field to change the lubricants or check the bolts on the wind turbines. "I don't enjoy the paper-pushing part of my job so much, but when you install a turbine and get it going, it is really kind of exciting. The result is a good, clean, pretty-looking wind turbine that runs good and churns out electricity when you're done."

Where do you see this job leading you?

Walter would like to see his company in Texas flourish. If that doesn't happen, he will try to get hired by one of the larger wind companies.

56. Energy-Efficiency Consultant

description: To put it simply, energy-efficiency consultants help clients save energy. They primarily do this by designing heating and cooling operations or evaluating existing systems in buildings and complexes. They are usually engineers by trade and are experts in energy conservation and energy uses. These efficiency experts conduct on-site energy audits, examining facilities and operations and looking for more cost-effective energy measures. This job goes beyond that of a simple energy audit. It may mean designing a new heating or cooling system or making recommendations regarding the use or refitting of equipment or machinery. It also includes overseeing installation and maintenance of the system. The job, with its engineering aspects, is much more technical than that of an energy auditor or weatherization expert, but it includes the skills of those jobs.

salary: For those just out of college, engineering salaries start in the mid- to high $20,000s.

prospects: Prospects for these jobs are tied to energy prices. If energy prices stay low, there's less incentive to improve efficiency, and therefore prospects in this field may be nominal. If energy prices go up, the demand for energy-efficient systems rises. On the other hand, there is always a demand for mechanical engineers to evaluate old systems and research new cost-saving energy technology, such as efficient lighting.

qualifications: An engineering degree is recommended, either in mechanical or electrical engineering. A physics background is also applicable. Weatherization work does not require a degree, but you need to understand the basics of heating and cooling principles and of electricity and thermodynamics. Good writing and research skills will give you a leg up on the competition.

characteristics: Energy-efficiency consultants rely not only on their technical expertise but also on their "people skills." As one puts it, "You must have the willingness to listen to a person who may not be as educated as you but might have the hands-on information about the facility that you're looking at." The job takes poise and tact and a bit of political finesse as you convince corporate clients of the necessary changes and modifications they must make.

Thomas Sahagian *is an energy-efficiency consultant.*

How did you get the job?

Thomas Sahagian started his career by getting a summer internship with an organization that engaged in wind power research in the Bronx. His job included maintaining the windmill on a daily basis and recording data. That job led to one with New York City's housing, preservation, and development department doing energy conservation work. As a technical services coordinator there he recommended energy-saving measures for multi-family dwellings. He checked things such as boiler efficiency and offered weatherization suggestions. That led to work with similar programs. On the job, he met employees of the engineering firm for which he now works, a firm that specializes in energy-efficiency consulting.

What do you do all day?

Thomas says he spends three or four days a week in the office and the rest of the time in the field meeting with clients and surveying operations. It's not merely a technical job. Thomas spends a lot of office time writing proposals, reports, and memos. Prospective clients will put out a call for estimates. After evaluating the site, Thomas himself will write the proposal in the hope that his firm will be hired to do the work. That's where good writing skills come into play. "It's extremely important to communicate information in a coherent and understandable fashion," Thomas says.

Some of his firm's biggest clients are institutions with large facilities and government buildings. Thomas also does work on residential buildings. For example, he's been hired to write specifications for new heating and air-conditioning systems. He also does a lot of troubleshooting. "A client may ask why they're using lots of energy. So we might set up local mechanical or electrical systems to see where there's excessive usage," Thomas says.

He and his colleagues also do what in the business is called "peer review"—that is, they are hired to review energy management systems designed by others. It's much like getting a second doctor's opinion. This is where "people" skills become as important as technical expertise. "You are immediately placed in conflict with someone who designed that system, and passing judgment. You're immediately plunged into what could be a very contentious, conflicting situation. So if you don't have the political skills to work around these obstacles, you could really compromise your effectiveness," Thomas says.

Where do see this job leading you?

Thomas wants to stick to the work he's doing now. "I gain satisfaction out of saving energy. I feel like I'm doing something for the environment."

> **WHILE IN COLLEGE, TRY TO GET A WORK-STUDY JOB WITH AN ENGINEERING FIRM. YOU MAY BE DOING ONLY "GO-FER" WORK, BUT IT WILL HELP YOU DEVELOP THE NECESSARY CONTACTS.**

57. Energy Auditor

description: Energy auditors are experts in energy conservation and energy uses and check for energy efficiency in homes and commercial businesses. You'll find these jobs with utilities. Most utilities offer this service free of charge to customers who phone for appointments.

On a typical home visit, the energy auditor first sits down with the homeowner, finds out his or her concerns, and examines recent utility bills. An audit requires getting to know the homeowner's use of electricity, which involves an in-depth interview: how many showers a day they take, how much they run air conditioners, whether they leave lights on when they're not home, how much laundry they do, and so on.

It's also a physical and sometimes dirty job, as the auditor then conducts a home inspection, climbing into attics or crawling under houses to check for insulation. An auditor usually makes two to three home inspections per day.

salary: On the West Coast, starting salaries for this job are $30,000 and higher, but that's above average compared to the rest of the country. Most municipalities offer starting salaries in the $20,000–$25,000 range.

prospects: The prospects for the job can be cyclical. A big expansion in the field occurred about five years ago. Now it seems that, for budget reasons, utilities are cutting back on these positions. However, to promote energy conservation, utilities are known to gear up marketing campaigns touting these free services.

qualifications: You need to have fundamental energy conservation knowledge that includes a basic understanding of heating and cooling principles. You must know your math, because you'll be doing lots of calculations when showing customers their billing history and usage of equipment.

characteristics: You have to be personable. An energy auditor is out there every day meeting customers in their homes. It takes poise and tact to show customers what they're doing wrong energy-wise and to convince them to make the changes necessary for an energy-efficient home.

Michael Zannakis *is an energy auditor for a large West Coast utility company.*

How did you get the job?

Michael Zannakis saw a newspaper advertisement announcing that the local utility was expanding its energy conservation department and looking for energy specialists. He got his training with the Australian air force doing energy conservation in remote areas of that country. In those sparsely populated, remote areas with limited utilities, the air force was educating the public on efficient use of energy.

Michael got married and moved to the West Coast of the United States, where he now works for one of the country's leading utility companies.

What do you do all day?

Michael travels with a laptop computer, asking very detailed, often personal questions to get an idea of how a homeowner or business is using energy. Every day it's a different house or business. One day he may find himself auditing a tiny home in a poor part of town in the morning, and a mansion in the afternoon. One morning, he says, he arrived at a house and an elderly woman greeted him at the door, insisting he first eat a lavish breakfast she'd prepared.

A home inspection involves getting up in the attic to check air-conditioning ducts. It also means checking crawl spaces under houses. Michael has seen his share of rats and spiders, and once even got attacked by a dog that didn't welcome the invasion of his domain under the house.

Michael also does commercial business energy audits, which take additional training because the equipment involved is larger. In these businesses you're examining large-scale machinery and different types of lighting than are traditionally found in homes. On such a visit, Michael might suggest different methods of operation or calculate what a business might save if it replaced a motor.

Michael has recently moved up the ladder to supervisor, so he's not going out in the field as much. But recently, he was headed out to check firsthand on a business that was having trouble curbing its energy use. The job requires diplomacy when a customer complains that he or she is not seeing savings on utility bills, while you know that the customer simply isn't following the suggestions you've outlined.

> **KEEP AN EYE OUT FOR NEWSPAPER ADVERTISEMENTS IN WHICH UTILITIES ANNOUNCE THAT THEY ARE CONDUCTING OPEN EXAMS FOR EMPLOYMENT. YOU SHOULD ALSO KEEP CURRENT ON ANY ARTICLES ABOUT EXPANSION OR CHANGES OF UTILITIES IN YOUR COMMUNITY.**

Where do you see this job leading you?

Michael wants to continue to move up the ladder into upper-management positions at the utility where he currently works.

58. Solar Power Inventor/Manufacturer of Alternative Energy Devices

description: Say the words "solar energy" and many people think of the large panels seen on some house rooftops. The science of solar energy has come a long way from these: there are now solar energy collectors that can fit in the palm of your hand. Manufacturers are finding new applications for solar energy every day and are creating new products—everything from rechargeable batteries to solar-powered, outdoor fluorescent lighting. These new power systems are creating possibilities for people to buy and develop remote properties that are removed from traditional power sources.

It takes someone to design and to invent this technology, and today, many utilities are hiring alternative energy specialists to design this new hardware.

salary: It's difficult to estimate what a self-employed manufacturer can earn because there are so many variables involved: creativity, entrepreneurial spirit, business savvy, etc. However, alternative energy specialists with public utilities can make salaries ranging from $28,000 to $30,000 a year.

prospects: The prospects are very good for this field as society looks for more efficient and cheaper energy. The demand by Third World countries for solar technology is especially responsible for this expansion.

qualifications: Mechanical expertise is essential. Designers of alternative energy systems for homes get practical training in everything from electrical engineering to plumbing. Theoretical training about solar energy and other alternative energy resources is recommended. You might want to get a degree in environmental science or one of the other sciences. One manufacturer says that he looks for people who not only have good knowledge about solar energy but also can actually solder wires together. Computer knowledge is also required.

characteristics: "You've got to be a little crazy to do this work," says one solar energy device manufacturer. It takes a curious, imaginative, and very creative person to invent and design these new products.

Bill Hollibaugh *is an inventor/manufacturer of alternative energy devices.*

How did you get the job?

Bill Hollibaugh became interested in solar energy when he was living in a remote part of northern California. He was making a living as an artist and sculptor when he came across a book that inspired him to design a device for heating hot water from wood stoves. He later sold more than 57,000 of these gadgets.

Although he took some detours into teaching and working in electronics along the way, he eventually opened his own business specializing in creating and manufacturing alternative energy devices.

> **GET AN IDEA AND GO FOR IT. THE MARKET IS AS WIDE AS YOUR IMAGINATION. ALSO KEEP IN MIND THAT MOST OF THESE JOBS ARE FOUND ON THE WEST COAST, ESPECIALLY IN CALIFORNIA.**

What do you do all day?

Bill's company produces more than 30 different products—everything from solar-powered water fountains to solar-powered vent fans used in outhouses in state parks. They also produce solar-powered fluorescent outdoor lighting fixtures that don't require extensive wiring. He just finished fine-tuning his latest design—a solar-powered water pump that can pump water 200 feet up, from a stream to a house on a hill.

On any given day, an architect may call asking Bill to design an entire system for a new home. Recently, Bill had just finished designing systems for two new homes. With the help of computer models, he estimated how much power would be needed and what hardware would be necessary for the jobs.

He just finished consulting on a movie which required unusual and intricate prop designs that were supposed to be alternative energy gadgets.

Bill also writes and publishes many books, some to accompany products and others to present general information about alternative energy.

Where do you see this job leading you?

Bill wants to continue to find new markets around the world, and to keep coming up with new inventions. Bill says there is plenty of opportunity in the field. He also sees potential for development in solar power advancement in, among other things, areas like water purifying systems, water pumping equipment, mobile hospital equipment, communications systems, and agricultural equipment.

59. Solar Energy Research Scientist

description: Although solar technology has been around for a number of years, it is a relatively new and young research field. Scientists are experimenting to develop new products. They are also searching for new ways to improve energy efficiency, reduce factory costs, and adapt solar energy for commercial use. Their innovations are already being put to work in telecommunications and in creating power for remote areas of the world.

salary: If you are lucky enough to be hired by a large corporation, the salaries are fairly good. Those with bachelor's degrees can start at $38,000, while those with a Ph.D. can earn a starting salary of $55,000 to $60,000.

prospects: This is a field on the cutting edge of technology and in which the research possibilities are endless, but the amount of emphasis companies place on new research depends on profits.

qualifications: Solar energy research scientists come from many educational backgrounds, including physics, chemistry, electrical engineering, chemical engineering, and mechanical engineering. There are research jobs open within all educational levels, from an associate's degree to a Ph.D.

characteristics: This career demands someone technically inclined, innovative, and analytical. He or she also has to be a team player; experiments are usually a collaborative effort among several researchers.

Li You Yang *is a solar energy research scientist.*

How did you get the job?

Li You Yang has a Ph.D. in physics, and worked on solar energy research for his thesis. Through his studies, he met people at the company where he's now employed. This solar energy products manufacturer was using some of the same materials that Li You was using in research. Li You simply gave the manufacturer a call and, with his extensive science education and experience, had no trouble getting a job.

What do you do all day?

Li You works primarily on photovoltaic cells of solar devices. These components turn solar power into electrical energy. He's researching new ways to fabricate the cells to improve energy efficiency. Ultimately, the company hopes to cut the cost of these products.

Li You oversees several experiments at once and supervises a team of researchers. He keeps a constant eye on on-going projects. It is a team effort to develop and test hypotheses, evaluate data, and plan the next step.

"We're in a constant cycle of designing experiments and evaluating results, making decisions, corrections, and improvements," he says.

The research requires spending a good amount of time at the computer.

> ONE SOLAR ENERGY RESEARCH SCIENTIST SAYS THAT WHATEVER LEVEL YOU'RE AT— WHETHER YOU HAVE A BACHELOR'S DEGREE OR A PH.D.—YOU MUST "DEMONSTRATE TECHNICAL BRILLIANCE."

Some of the days are long. A critical point of an experiment might occur just at 5 o'clock, causing him to have to stay late.

Where do you see this job leading you?

Before taking the job eight years ago, Li You had the choice either to teach or to head into research. He's happy with the decision he made, and wants to stick with alternative energy research.

60. Agricultural Horticulturist

description: An agricultural horticulturist applies the science of raising plants to the business of running a farm. In this line of work, you're looking at how much work and money it'll take a farmer to raise a salable crop. Potential employers include county agricultural extension offices, nonprofit research foundations, private farms, nurseries, and seed companies.

salary: Entry-level horticulturists can earn $20,000 to $30,000 per year.

prospects: A county extension office provides the easiest entry-level route into this field. As you gain technical knowledge and make professional contacts, you may move to a nonprofit research foundation or to an agriculture-related support business.

qualifications: A degree in horticulture or a related scientific area, along with some farm experience, is desirable.

characteristics: The most important quality an agricultural horticulturist must have is an understanding of how difficult a farmer's job is. All of a horticulturist's science is geared to help the farmer keep his or her head above water and make a profit. You should be able to listen as well as instruct, since each farmer's problem will be slightly different.

Alan Ware *is a horticulturist with a sustainable agriculture center in Poteau, Oklahoma.*

- -

How did you get the job?

Alan Ware spent his first three years after college managing a large blueberry farm. When he saw a chemical accident nearly claim the life of a farmworker, he vowed to stop working in close proximity to toxic chemicals. The job he has now was advertised in a professional horticulture journal.

Alan has a bachelor's degree in horticulture. His first employment choice was to work in greenhouse plant production, but the greenhouse industry was "in a mess," he says, when he graduated, so he took the blueberry farm job.

> TO HAVE A GOOD CAREER IN THIS FIELD, YOU SHOULD GET SOME TECHNICAL TRAINING. CONSIDER SPECIALIZING IN BIOTECHNOLOGY, AN AREA OF STUDY THAT CONCENTRATES ON PLANT BREEDING; SOCIETY WILL CONTINUE TO SEARCH FOR MORE NUTRIENT-RICH VEGETABLES OR INSECT-RESISTANT PLANTS.

What do you do all day?

Typically, Alan spends a couple of hours a day on one of the demonstration farms at the center where he works. There they raise and test many different crops and bring in farmers to see how the same plants might work on their property. The demonstration farms grow strawberries, blackberries, shiitake mushrooms, vegetables, ornamental trees, bedding plants, and more.

Recently, Alan was doing a lot of phone work, arranging for farm visits, and working on an article for a bimonthly newsletter. He gives talks to area garden clubs and farmers' market groups. There is also some research involved, in cooperation with local universities.

Alan says he introduced the idea of growing shiitake mushrooms to some small farmers in his area because of the low capital outlay required and the excellent profit margin. On his visits to area farms, he meets with farmers to discuss production goals. Alan is ready to give advice regarding elements of farm management such as costs, budget, and expenses. Sometimes, he says, farmers need very basic advice on what crops to grow and how much. But helping farmers grow crops is just one part of the work Alan does.

He also encourages them to discover and target new markets. Many of those he advises are part-time farmers with less than 20 acres. "We can see progress. We keep files on every farmer and can track progress through the years," Alan says.

April through July is the busiest time because he gets so many requests to visit farms, and the demonstration schedule is just beginning. In the fall, he leads education programs and workshops for the farmers. Winter months are quiet.

Where do you see this job leading you?

Alan loves his job and where he lives. "I wouldn't leave here if they fired me three times and kicked me out the door," he insists. He would, however, like to own his own business someday—maybe a landscaping firm, a nursery, or even a farm.

61. Horticulturist

(PUBLIC GARDEN)

description: Horticulturists are responsible for the maintenance and upkeep of botanical gardens. Their responsibilities might also include the greenhouse and conservatory. Horticulturists/curators design spaces, order seedlings, and are responsible for ordering supplies, controlling pests, and maintaining tools and equipment—all the things necessary for a beautiful garden to flourish.

Botanical gardens and arboretums teach people about plants and provide refuge for rare and endangered plants. They also contribute beautiful respites of splendor and tranquility in the midst of harsh, concrete-filled cities.

salary: A horticulturist for a public garden can earn anywhere from $15,000 to $30,000 per year.

prospects: Jobs with public gardens can be limited in number because most communities have only one public botanical garden. Many botanical gardens also operate at the whim of public funding or charity fund-raising.

qualifications: It takes more than just a "green thumb" to make it as a horticulturist at a major public garden. You don't need a college degree to work in a public garden, but you'll need that education if you want to advance and make a worthwhile career of the job. Many of the better jobs require degrees in plant sciences such as botany, horticulture, forestry, or biology. A master's degree will help you attain management or research positions.

characteristics: Creativity is the key to becoming a successful horticulturist. You also have to be able to think on your feet and interact with the public. You have to love the outdoors and tolerate working in all kinds of weather. To do this job well, you need to be in good physical condition.

Cheryl Lowe *is a horticulturist for a flower society in New England.*

How did you get your job?

For as long as she can remember, Cheryl Lowe has loved the outdoors and, in particular, plants. She recalls growing up wanting to be a naturalist. After college, she worked with the Nature Conservancy and for a local university and local parks department. She soon went back to school and earned a master's degree in public horticulture administration. Cheryl saw an advertisement for her current job in a professional magazine published by the American Association of Botanical Gardens and Arboreta. She is the director of horticulture for the New England Wild Flower Society, which operates a public garden.

What do you do all day?

A typical day depends on the season. Spring is the busiest time of year. It's showtime as the staff spruces up the garden to open to the public. That's when the horticulturists get down on their hands and knees to clean up beds, pick up leaves, and push dead tree limbs

> VOLUNTEER AT A PUBLIC GARDEN. CONTACT THE AMERICAN ASSOCIATION OF BOTANICAL GARDENS AND ARBORETA (AABGA) FOR A LISTING OF PUBLIC GARDENS. ASK STAFFERS AT A PUBLIC GARDEN IN YOUR AREA FOR NAMES OF LOCAL AND REGIONAL PLANT SOCIETIES.

through the shredder. Cheryl and her crew take care of some 1,600 different kinds of plants in the 10-acre garden.

The horticulturists' work is never-ending. On a recent hot summer day, Cheryl had spent the morning weeding the garden. They were readying for a camera crew from *Good Morning America* to come the next day to tape a segment on wildflowers.

Being a horticulturist is much more than just planting seeds and watching the garden grow. It can be backbreaking work. Some days Cheryl cuts logs with a chain saw to repair trails. Larger gardens, such as New York's Botanical Garden, often have teams of horticulturists and maintenance workers, but at this smaller garden, Cheryl finds herself changing

light bulbs, shoveling walks, and even cleaning toilets, in addition to her horticultural duties.

She and another horticulturist and propagator who works in the nursery during the winter decide which new plants and seeds to order. They also work with a special rare and endangered plants division. Cheryl says her office is the outdoors and she loves the way the office walls change with the seasons.

Where do you see this job leading you?

Cheryl can get a job working in a larger public garden, where she wouldn't have to do the building maintenance tasks she does now. Right now, she's busy creating new projects such as renovating a meadow. Overall, she finds the work rewarding. "I think these gardens make a difference. It's not a huge difference, but it makes an impact on how people see their immediate environment."

62. Ecological Horticulturist

description: An ecological horticulturist is a scientist with an environmentalist's point of view. In this field, you will apply various soil and plant sciences to the management and raising of crops, by using products and practices that are both economically feasible and environmentally sound. There are a few alternative-agriculture institutes that hire such researchers. Some of these horticulturists also work as consultants—for instance, helping farmers protect their crops from pests without, or with limited use of, toxic chemicals.

salary: A graduate with a master's degree may start as a technical assistant working for a research project, in sales, or in consulting. Salaries in those jobs start at around $25,000 to $30,000.

prospects: There is a growing demand for ecological horticulturists within universities and with the U.S. Department of Agriculture. A good way to make contact with one of the established or upcoming nonprofit organizations in this relatively new field is to work through agricultural cooperatives, with organic farmers, or with faculty members at various colleges and universities. The ongoing interest in so-called "health" foods and environmentally responsible commercial products makes this a growing area of study.

qualifications: A master's degree offers more flexibility in the job market and a higher salary. A Ph.D. is recommended if you are truly driven to break new ground in this research and want to supervise your own projects.

characteristics: You will need a sense of mission as well as an understanding of the sciences that drive agriculture: soil management, horticulture, genetics, and biology.

Terry Schettini *is an ecological research horticulturist at a private institute.*

- -

How did you get the job?

Terry Schettini got the bug to work in alternative agriculture while attending high school. His chosen area of study was genetics and plant breeding, because, at the time (early 1970s), it was the best option for looking into ways of growing food without synthetic chemicals. By the time he earned his Ph.D., he was a firm believer in "regenerative" agriculture, or the need to replenish natural resources while growing food.

Terry had been reading the magazine *Organic Gardening* since his high school days and was aware that its publisher, Rodale Press, had a sister organization (the nonprofit Rodale Institute) dedicated to the same principles he'd embraced during his studies. He worked with several chemical companies just before he actually landed a job at Rodale Institute. He was worried that his work on the "other side of the fence" might put off his prospective employers, but they seemed to think it gave him a good background in understanding why farmers use pesticides in the first place. Rodale has long been a mecca for people interested in alternative agriculture. Terry landed his first job there as horticultural researcher and has worked with vegetable and apple production systems, as well as with communities interested in supporting healthy agriculture. "I consider it a luxury," he tells us. "I get paid to do what I had to do after hours before."

Terry was surprised recently to see a job similar to the one he first landed at Rodale advertised in a professional journal. He says it shows that the idea of regenerative agriculture is becoming more popular.

What do you do all day?

During the growing season, Terry's days are spent on one of the institute's research farms, supervising plant and soil samples. Trials are now under way comparing conventional fertilizers to organic compounds. Four separate crops are being evaluated in terms of nutrients, taste, and shelf life of the product.

Winter is a more stressful time. That's when many grant proposals are due and Terry must finalize department budget decisions. This past season he spent a lot of time organizing a conference and planning for the tour groups that visit the experimental farm.

Where do you see this job leading you?

Terry seems to be where he wants to be. He's trying to answer the question "Is organic food really healthier for you than the products of conventional agriculture?" He has his instinctive opinion, but as a scientist he wants to be able to prove the concept to others. Terry's a firm believer in the institute's mission: healthy soil, healthy food, healthy people.

> **TO FIND THE APPROPRIATE EDUCATIONAL PROGRAM, IT'S SUGGESTED YOU INTERVIEW RECENT GRADUATES IN THE FIELD.**
> - - - - - - - - - -

63. Nursery Plant Manager/Owner

description: It takes more than just having a "green thumb" to run a successful nursery. To succeed, a nursery manager needs a horticultural background, business savvy, and the resilience to bounce back from failures sometimes caused by factors totally out of his or her control—such as bad weather. A manager's responsibilities may include planning overall production, selecting and ordering seeds from catalogs, and purchasing cuttings from other nurseries. You may manage a nursery with one greenhouse or dozens. There may not be a more fruitful yet delicate business than that of horticultural nurseries. As a nursery manager, you might find yourself awake in the middle of a winter night watching a thermometer for impending freezing temperatures that threaten to wipe out a year's worth of work and profits in a few short hours. A nursery plant manager must nurture not only the greenery, shrubs, flowers, and vegetables, but also the personnel. In larger nurseries the manager may oversee a staff that includes plant breeders, propagators, or horticultural engineers as well as maintenance workers. You may be taking care of everything from personnel schedules and payroll to purchasing. Whether you're an entry-level propagator or a top supervisor, count on getting your hands dirty.

salary: Entry-level positions start at minimum wage, but top pay for a supervisor or manager may hit the $42,000 level.

prospects: It's growing! You could say the nursery business is beginning to blossom once again. It's just now bouncing back from the recession of the 1980s. There's new opportunity with the offshoot of new specialized nurseries emerging on the market. Look for nurseries that, for example, offer hydroponics propagation or that grow native plants and grasses to restore woodlands and wetlands.

qualifications: A degree is helpful but not mandatory. In the past, most experienced nursery workers learned everything on the job. But now nursery owners are looking for well-educated horticulturists with a knowledge of botany.

characteristics: This business demands creativity, patience, and tenacity.

Nancy Vermeulen *is the owner of a medium-size East Coast nursery.*

How did you get the job?

Along with her brother, Nancy Vermeulen co-owns and manages the nursery her grandfather started 75 years ago. Growing up, Nancy never intended to one day take over the family business. She loves animals more than plants and wanted to become a veterinarian. She worked as a veterinarian's assistant. But when her father's secretary became ill, she stepped in to help at the nursery. As she says, that's when she realized, "These are my roots (pardon the pun); this is my heritage."

Their wholesale nursery starts plants from seeds, root cuttings, and grafting. They shepherd plants from sprout to maturity, starting the seedlings in greenhouses and eventually transferring them to outdoor container areas. The Vermeulens don't grow houseplants. Instead, their nursery stock con-

> **WORK SUMMERS IN CROP PRODUCTION OR PLANT MAINTENANCE. VOLUNTEER AT THE LOCAL BOTANICAL GARDENS. YOU CAN GET A JOB AT A NURSERY RIGHT OUT OF HIGH SCHOOL, BUT IF YOU WANT TO ADVANCE OR IF YOU DREAM OF ONE DAY OWNING YOUR OWN NURSERY, TRAINING IN HORTICULTURE, BOTANY, AND BUSINESS IS REQUIRED.**

sists of landscape shrubs and trees, and they specialize in bonsais. As wholesalers, the Vermeulens sell mostly to other nurseries and landscapers nationwide.

What do you do all day?

"First thing I do is make the rounds checking how propagation is going. This is when we catch up on what needs to be done, like potting or general maintenance of the entire operation."

September is when the nursery starts planting again. It then gets busy straight through the winter with propagation and grafting. In March, as Nancy puts it, "All hell breaks loose in the nursery business" as they reap the benefits of what they sow, shipping plants, trees, and shrubs to clients nationwide. In June and July they recoup, go to trade shows, and comb seed catalogs as they plan the following year's ventures.

Nancy does spend time in the office. She manages 15 employees and has to stay on top of administrative work, as well as take care of marketing, sales, and cultivating new clients. But she says that she tries to spend most of her time out of the office getting her hands in the dirt.

One of the toughest times the business had was two years ago, when New Jersey, along

with much of the East Coast, experienced record snowfall. Through 17 snowstorms, she and her brother struggled to protect greenhouses and outside plantings. It was an endeavor just to uncover the plants through the snow. Though ice storms destroyed crops and the weight of heavy snowfall crashed through the greenhouse roofs of many area nurseries, the Vermeulens didn't lose one greenhouse. That's because her brother stayed out all night working to ensure that it wouldn't happen.

Where do you see this job leading you?

Nancy looks toward expanding the nursery. She is constantly searching the market for rare and unusual plants. "The challenge is finding the profitable plants," she says. "As much of a struggle as it is, we love what we do."

64. Plant Health Care Worker

description: Working in the industry of plant health care doesn't necessarily mean you will be conducting routine plant maintenance day in and day out. More than likely, you will be brought onto a job only during a crisis situation. It will be necessary for you to come up with solutions to situations that range from a persistent insect problem to adjusting the pH levels in a yard of soil. You will work with both indoor and outdoor plants, but your largest, and perhaps most complicated, jobs will require that you handle problems on a large plot such as a golf course, botanical garden, or large estate.

salary: Starting salaries for this type of work are usually in the area of $28,000 a year.

prospects: Finding a job in this field could be tricky because it is so new and specialized. However, the plant health care industry is growing, and with it the number of jobs available.

qualifications: For this type of work, it would be helpful to have a background in the sciences, such as botany or chemistry. But perhaps more important would be to gain your certification as an arborist, which would give you specialized knowledge in plant disease.

characteristics: People working in plant health care have compassion not just for animals and humans but for every living thing. People in this field enjoy being outdoors and love working with plants.

Ron McCulty *is the owner of a plant health care company.*

How did you get the job?

During Ron McCulty's senior year of college, he received an offer to become the sales director of a yard care/chemicals manufacturing company. At the time, Ron says, the $70,000-a-year offer was too good to pass up, so he left college to begin his career in chemical sales. However, all the while Ron was working in this industry, he felt remorse for the effect his work was having on the environment. "I remembered how my mother used to help me press the leaves into wax for biology. She was a lover of nature, and it was the memory of her that really got me to the point where I just decided that I had done so much to hurt the environment, that I was going to spend the balance of my life paying the environment back." Ron left the chemical industry and started his own plant care company, which uses mostly homeopathic remedies for complex yard problems and thus avoids the use of harmful chemicals.

What do you do all day?

For the most part, Ron says, he spends his day solving problems. However, the range of problems he finds himself solving from day to day varies greatly, and he says he has yet to come across two that are alike. "Never are the problems the same. I've never seen a property that is even closely related to another property that we've done, because everybody has different taste. Some people like azaleas, some people like fir

> PEOPLE LOOKING FOR WORK IN THIS FIELD WOULD BE WISE TO GET EXPERIENCE WORKING WITH PLANTS WHILE GAINING THEIR EDUCATION. A NURSERY OR BOTANICAL GARDEN IS A GOOD PLACE TO FIND A SUMMER JOB OR VOLUNTEER YOUR TIME.

trees, some people like rocks, so you have to have this extraordinary range of capabilities," Ron says.

Aside from the paperwork and phone time that go along with running a business, Ron spends a large part of his week working outdoors. He may be requested to rid a property of pesky ants, or he may be employed to save a lot full of rare trees that have been planted incorrectly and are dying because of it. No matter what type of task he is performing, he always keeps in mind how his solution to a problem will play out in the environment.

Where do you see this job leading you?

Ron says he has reached the pinnacle of his career and doesn't foresee making any career changes. "Literally, I feel like I was put on this planet to do what I do now," he says.

65. Organic Farmer

description: Sales of organic food in the United States are growing by more than 20 percent each year. Not only are more individual commercial farmers venturing into organic farming, but larger companies are doing so as well.

Organic farming requires a special knowledge and education in plant and soil sciences because these farmers are producing crops without the use of chemical fertilizers or pesticides. To be successful, organic farming requires frequent analysis of soil and plant tissue to ensure that the crops are rich in nutrients. Being an organic farmer or farm manager involves all the responsibilities of traditional farming—from seed and crop selection to cultivation and harvesting, driving tractors, and running a business.

Today, organic fruits and vegetables are becoming a part of our mainstream diet and are being sold at neighborhood grocery stores. Large produce companies are jumping on the "organic" bandwagon, growing these crops not only to be sold as fresh fruits and vegetables, but also for canning and use in gourmet food products.

salary: It's difficult to estimate what an organic farmer will earn, since this depends on a combination of factors: acres, crops, market, business savvy, and luck. However, those working as farm managers—overseeing hundreds of acres for a large company—may earn $30,000–$40,000 per year.

prospects: The organic food industry is expanding every year, and there is demand for those who are educated in this specialized agriculture.

qualifications: A degree in agriculture, botany, soil science, or environmental science is useful, though the best education is to work alongside an organic farmer, seeing firsthand how organic crops are cultivated, problems are tackled, and solutions are created.

characteristics: Just as in running any small business, the job of an organic farmer requires expertise in several specialties: not only agriculture but also biology, botany, soil science, marketing, and sales. The work takes stamina; days begin before dawn and the physical labor is strenuous. It also takes instinct. "Farming is problematic because it does require a real instinctive understanding of who and what you're working with. It's not like other businesses where there's a clear way of doing things," says one organic farmer.

Mark Lipson *is an organic farmer.*

How did you get the job?

While in school earning an environmental planning degree and studying farm cooperatives, Mark Lipson and a group of friends combined their financial resources to buy a farm together. They started by purchasing four acres. Thirteen years later, they own 137 acres, of which they farm about 30. They primarily grow organic vegetables, tomatoes, and squash.

Although none of the 13 adult partners had agriculture degrees, some came from farming backgrounds and knew about equipment like tractors. They pooled their knowledge and have created a thriving organic farming business.

What do you do all day?

Workdays begin before sunrise. During harvesting season, Mark starts by checking crops and determining what he's going to have to sell that day. He instructs the hired crews or field supervisor about the harvests. He then makes sales calls and determines orders. During the day, Mark checks the quality of what's being harvested. Since there are always crops at different stages of growth, there's always work to do. The farmers plant year-round, so on any given day they may be preparing the land, planting, or cultivating crops.

Being "organic" is more than just growing crops without pesticides. Mark says the essence of this special farming is maintaining the biological soil fertility. He and his co-workers must analyze what's going on in the soil by doing soil sampling and analysis. They also analyze plant tissue samples, and from these are able to determine if there is a nutrient deficiency.

Though the stress accompanying this work is certainly different than that caused by sitting behind a desk all day, Mark says there are pressures that come with the job. There are many details to keep up with and factors that are out of your hands . . . like weather.

Where do you see this job leading you?

Mark and his partners have built their own individual homes on the land and enjoy the lifestyle that this brings. Mark says that after more than 10 years as organic farmers, they are just now reaching the point where they are each earning $25,000 to $30,000 a year.

> **THERE ARE APPRENTICESHIPS AVAILABLE IN WHICH YOU CAN WORK ON ORGANIC FARMS OR COOPERATIVES.**

66. Organic Farm Inspector

description: In the past, organic food customers have relied on the good faith of growers and merchants to assure that what is being sold as "organic" is indeed organic—that is, food that is grown without harmful pesticides or chemical fertilizers. The organic food industry is largely unregulated, but that is changing; some states have enacted labeling laws and have hired private certifying companies to verify that the food is organic. These certifying companies employ farm inspectors to test soil samples and check growers' purchase records—all to confirm that the farmers are upholding organic practices. There are also inspectors who check processes rather than the fields. These specialists examine packing facilities to make sure organic foods are not mixed with conventionally grown foods. They also check all the systems in which organic foods are processed into other products, like sauces.

Most organic food inspectors are independents hired by certification companies. Many are farmers doing this work on a part-time basis in off-seasons. But there has been a shift to hiring those with broader professional agricultural backgrounds on a full-time basis.

salary: For those working on a part-time basis, the hourly wages start at $14. Those with more experience can start at $20 an hour. Full-time yearly salaries are about $25,000.

prospects: A federal law is pending that will regulate label requirements for organic fruits, vegetables, meats, and processed foods. If this passes, certification for organic farms may become mandatory, thus creating a higher demand for inspectors.

qualifications: While an agricultural background is obviously needed, you may find more organic education being taught in horticulture courses. Writing skills are important; the reports written by the inspectors are what committees use to make their decision on each farm's certification. It also helps to know Spanish, especially if you're working on the West Coast, where many new farmers are Mexican-Americans. The Independent Organic Inspectors Association offers training in this field.

characteristics: Being a "morning person" is important. The job usually starts between seven and eight in the morning, by which time the farmer is on his third cup of coffee. Inspectors should also be punctual. Showing up on time is very important in this business.

John Foster *is an organic farm inspector.*

How did you get the job?

John Foster has a broad academic background, with degrees in agriculture, horticulture, environmental studies, and American literature. While in school, he worked for a short time on an organic farm, but after graduation, he worked mostly in conventional agriculture, where he essentially advised farmers on which pesticides to use. Philosophically, he didn't like what he saw when it came to the use of chemical pesticides and was thinking about a change when he heard about the field of organic farm inspecting. He decided to call a certification company and ask for a job. As luck would have it, they had an opening. Although he hadn't been trained as an inspector, John's extensive agricultural

ONE OF THE MANY CONFERENCES ON THIS TOPIC IS CALLED ECO-FARMS AND IS HELD BY THE COMMITTEE FOR SUSTAINABLE AGRICULTURE.

background and experience fit the bill. He received training on the job, and also takes annual refresher courses.

What do you do all day?

John's up early. He usually starts his meeting with a farmer at around seven or eight in the morning. "I primarily am an observer and recorder," he says. He inspects every farm according to three general areas: farm records, equipment and materials, and the fields. His job is to verify that the organic crops are

organic—meaning that they are being cultivated and grown according to organic standards.

Since most of the farms he inspects consist of less than 200 acres, each inspection takes less than a day. He starts by checking the farmer's records, which is the most difficult part of the job. It's analogous to a tax audit. "Each farmer has his own record-keeping system . . . one may keep records in a milk crate, another in a file cabinet."

He also does equipment checks. "I see how they take care of the weeds, if they have the equipment to do it in the manner that they told me they do it. In orchards or vineyards, if I see a very clean, weed-free field, that raises a flag that they may not be [using] totally organic methods."

He then checks the crops themselves. "I try to do inspections when they're harvesting or packing, to see how the crop goes from the fields to pack-

ages." John makes sure that even when it comes to trucking the harvest, there's no chance of the organic foods mixing with conventional products.

Back at the office, he writes up reports. These are very detailed, and critical to the farm's getting certification as a producer of organic goods.

Where do you see this job leading you?

John juggles two other jobs along with being an organic farm inspector. He works as a staff research associate for a university cooperative extension department and as a crop advisor for another college's horticulture department. He hopes to work full-time in the organic industry, and ultimately to work full-time consulting for farmers about organic farming methods.

67. Social Research Analyst

(INVESTMENT FUND)

description: Socially responsible investment funds are a relatively new trend in the financial investment world. These funds are designed for those who choose to invest their money in funds that meet a prescribed code of moral, ethical, and political criteria. Several brokerage houses and investment groups now specialize in counseling investors in these particular stocks, bonds, and money market plans. It's the job of the social research analyst to screen each of the companies considered for one of these funds. The social research analyst will check companies to find out their record on issues such as human rights or protecting the environment.

salary: Salaries start at around $30,000. The average salary for most experienced analysts is about $40,000 a year. Some nonprofit groups also hire research analysts; their starting salaries are around $20,000.

prospects: It's an emerging new career. More and more banks, brokerage houses, and others who manage assets, such as pension funds, have research departments to conduct social screening.

qualifications: A business or finance degree is not required, but is certainly helpful. A liberal arts background is acceptable as long as you possess good research, writing, and communications skills. Previous experience with a specific social issue or in corporate research is a must for getting jobs with investment companies, but not necessarily critical for nonprofit groups.

characteristics: This job requires a little detective work, so it takes a bit of an inquisitive nature. You also often have to be creative to get the necessary answers. However, the job can also be tedious. The methodology used to research companies is similar for each company, which can become repetitive.

Ken Scott *is an environmental and social analyst for an investment fund.*

How did you get the job?

Ken Scott is currently employed as a social research analyst at the Calvert Group, a mutual fund company and one of the nation's leading socially responsible investment funds. Interestingly enough, though, Ken doesn't come from a financial background. He says that after graduating with a degree in political science, he became interested in the environment when he read the book *Shopping for a Better World: A Quick and Easy Guide to Socially Responsible Investing*. The book so interested Ken that he moved to New York to take an internship with the organization that publishes it, the Council on Economic Priorities. That's where he began doing research and writing reports on companies' environmental records. The internship turned into a staff job. Investment houses frequently called Ken for information research, and eventually Ken was hired away by one of those investment firms.

What do you do all day?

Ken spends the day doing investigative research. Specifically, he examines companies' environmental records and reputations. Some companies, he says, take only a few hours to screen, while others may take days or weeks. Much of it depends on the size of the firm being checked out and the availability of information.

The job involves reading and evaluating volumes of paperwork, and working for hours on the computer. Ken must first sift through a company's annual reports and other related documents. He checks everything from a company's pollution control measures or emissions, to its holdings and management practices. Using computerized information services, he digs for any published magazine, newspaper, or journal articles. That's followed by searches of databases of government agencies, such as those of the EPA, to check things like pollution emissions. He also retrieves information from independent research groups and environmental groups.

Ken's job not only involves collecting the data—it also requires that he understand, interpret, and evaluate the information so he can make judgment calls and offer recommendations regarding a company's environmental record. One of the things he likes best about the job is knowing that sometimes his inquiry can influence a company's policies or disclosure.

IN ADDITION TO CALLING THE PERSONNEL DEPARTMENT OF ONE OF THE SOCIALLY RESPONSIBLE FUNDS, CALL THE RESEARCH DEPARTMENT DIRECTLY AND ASK THEM ABOUT JOB OPENINGS.

Where do you see this job leading you?

Ken says he's happy focusing on environmental issues. One of the daily challenges, he adds, is managing the needs of portfolio managers—completing research, within time constraints, that is comprehensive and consistent with past decisions. One of his goals is to convince companies to disclose more information about themselves.

68. Financial Environmental Consultant

description: A consultant gives expert or professional advice and helps clients solve problems. A financial environmental consultant advises clients on the financial ramifications of their environment-related business decisions. Overall, these consultants help with strategic planning, diversification, and identification of markets. A client may need help deciding to diversify into a new environment-related technology or to expand current products into a new market. A financial environmental consultant can also help a company to commercialize new technologies.

salary: Those with bachelor's degrees usually start at around $25,000 to $28,000, while those with graduate degrees start at slightly higher levels.

prospects: This is a very new career field. It is one that someone already working in the financial world or environmental industry might want to transfer into. Because there are very few independent financial environmental consulting companies, there is room for more.

qualifications: Technical and/or science degrees are required. One environmental consultant tells us that although environmental consultants are usually considered generalists, they need to be specifically trained in one area. She says that often the principles, methods, and disciplines you learn in one scientific area are applicable to others. This career also requires continuous education and training, and extensive reading of technical trade journals to keep up with changes in industry and science.

characteristics: Being able to work on a team is probably the most important skill an environmental consultant needs. Most often, you're part of a working team of consultants and advisers. An environmental consultant needs to be able to assess, investigate, research, and evaluate projects, and to then communicate even the most complex findings in language any client can understand.

Joan Berkowitz *is a financial environmental consultant.*

How did you get the job?

Joan Berkowitz is an outstanding example of someone who has been able to capitalize on her knowledge and experience and adapt it to changing environmental concerns. Joan has a Ph.D. in physical chemistry, and started her career with a nationally known consulting company. She worked on high-temperature chemistry related to solid rocket propellants and rocket nozzles.

When that field declined, she moved to another area of consulting—hazardous waste. In 1972, she helped the EPA write its first report to Congress on hazardous waste. She eventually was assigned to examine and revamp the hazardous-waste handling system at a Mississippi tannery.

It was while taking a nine-week senior executive course at MIT that, she says, she became

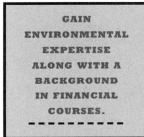

GAIN ENVIRONMENTAL EXPERTISE ALONG WITH A BACKGROUND IN FINANCIAL COURSES.

interested in the financial end of the industry. Today she and a business partner have created their own financial environmental consulting firm. They advise companies on the financial ramifications of making certain investments or technological changes. "Part of our current practice is in helping firms commercialize technologies and look objectively at the market for their technology," says Joan.

What do you do all day?

"We look upon the environment or the aspects of it as a business," Joan says. She spends most of the day on the phone talking with clients and their

competitors. They share views about the marketplace and research the financial and political climate. To stay ahead of the game, Joan reads some 25 journals every week. She also works on a yearly "state of the environmental industry" report, which is presented at environmental business conferences. Recently, she was compiling notes for a training session she was conducting regarding overseas environmental markets. Clients often call with tough questions that require research and complex answers. Joan examines each issue from a technical and financial perspective.

On any given day, she may get a call from a company wondering whether it should jump into the remediation market in western Europe or whether to purchase a particular technology. One client may wonder if it's economically advantageous

to license a new solid-waste treatment system, while another wants advice on a particular environmental investment or how to diversify. Yet another client wants to know how costly it would be to get out of the landfill business and into something else.

"Being ahead of your clients is what this business is about," Joan says. Her work demands that she keep in touch with environmental experts around the country. For that reason, she attends many conferences and is in big demand as a speaker.

Where do you see this job leading you?

Joan and her business partner are hoping for continued success and an endless list of clients and work. Her advice to those thinking about a career in financial environmental consulting is to always be in the process of re-educating yourself. Remember, she says, "The primary value of a university education is that you learn to learn."

69. Environmental Attorney

description: An environmental attorney is a lawyer with a specialty and a passion. It is not a lucrative field by legal standards, so the person entering this job needs to have a high level of commitment and belief in the work. You must integrate knowledge of the underlying science involved in environmental issues with expertise in using the legal system, as well as understand the way corporations do business. Generally, this is a litigation job. You will file suits and try cases. This is an active, creative, and sometimes frustrating line of work.

salary: Public interest groups start their attorneys at $21,000 to $31,000 annually, or about half what the same person would likely be paid at a private law firm.

prospects: There are a variety of nonprofit public interest groups around the country, and they have high turnover among their lawyers, so the prospects, in that sense, are good. On the other hand, it is difficult to get one of these jobs right out of law school.

qualifications: A law school degree and admission to the bar are required for this work. You will also need some schooling or a clear understanding of the underlying science for these types of cases. Previous litigation experience is often a necessary prerequisite. An environmental litigator also needs to be skilled in dealing with the media and the public, as well as in negotiating with corporations and polluters during the litigation process.

characteristics: In this game, you have to be able to juggle a number of different tasks simultaneously and then quickly focus on the challenge at hand. You need a "stomach" for litigation—that is, you must have enough of a contentious side to enjoy, on some level, the process of ordered conflict that is the American legal system. It's essential that you be able to write well and express yourself concisely and creatively. Frustration comes with the territory, so you must also be resilient.

Charles "Chuck" Caldart *is an environmental lawyer and litigation director at a nonprofit environmental legal center.*

How did you get the job?

Chuck Caldart was seven years out of law school, working at a private firm, when he decided to go back to school to gain the expertise needed for this specialty. At the Harvard School of Public Health, he learned the basic science that would form the core of environmental cases he would later take to courts from coast to coast. In his first job, he handled everything from child custody disputes, to medical malpractice cases, to business litigation. He got the environmental bug when he represented demonstrators accused of trespassing at the construction site of a nuclear power plant. He wrote a referendum on nuclear power that was passed by voters in the state of Washington, went back to school, and eventually took a teaching position at MIT. A desire to return to environmental advocacy led him to his current job.

> WHILE IN LAW SCHOOL, TRY TO WORK WITH A PUBLIC INTEREST GROUP DOING ENVIRONMENTAL WORK. BEGIN MAKING CONTACTS AND GATHERING AN UNDERSTANDING OF THE FIELD SO YOU CAN TAKE ADVANTAGE OF THE OPPORTUNITY WHEN A POSITION BECOMES VACANT AT A LATER DATE. REMEMBER, BECAUSE OF THE RELATIVELY LOW LEVEL OF PAY IN THIS JOB, THERE IS A HIGH TURNOVER RATE. YOU'LL GET YOUR CHANCE. BE PATIENT.

What do you do all day?

There is a lot of traveling. In the past five years, Chuck has litigated numerous cases in the Midwest and on the West Coast. Recently, he was about to leave for San Francisco to argue a motion, while two of his co-workers were headed to New Orleans to research a case. Chuck says he tries to schedule his travel periods in advance, but cautions that you need to spend a lot of time on the road. "Get used to airline food," he advises.

Cases usually come to the attention of the legal center in two ways. "It might stand out in a routine check of files as being particularly egregious, or someone in the public may call us with a lead asking us to check out something," Chuck says. The center must then determine how bad the polluter's actions are and how or where the legal team would have the largest impact. Chuck says the job also requires having a good business sense to pick the right cases and use your resources with maximum impact and benefit.

Cases fall into several different categories. In some cases you get immediate results. A simple notice of intent to file legal action prompts some companies to clean up their act. "However, there are other ones that are really fought and you're in court all the time," Chuck says. "Sometimes the case doesn't settle until just before the trial date . . . after a lot of hard work and time." As with all litigation, there are peaks and valleys: relatively calm periods of initial research and frenzied times of trial preparation that can involve all-night work sessions.

Where do you see this job leading you?

Chuck is happy with what he's doing; if he weren't, he would have stayed in private practice, where he made more money. Some of his former co-workers have gone back to the private sector, so that's always a possible career path. Chuck is guardedly optimistic that the opportunities and effects of the type of work he does will continue to expand into the next century.

70. Environmental Journalist/Editor

description: An environmental journalist is a reporter with a specific assignment: the environment is his or her beat. While some stories—such as oil spills and other ecological disasters—may be obvious, the real skill in this line of work involves conceptualization: the ability to look at a field of expertise and present news about it in a way that will interest the layperson. The environmental journalist will generally make his or her own assignments, "sell" them to the editor, and research and write the articles. You will usually have less editorial supervision than a general assignment journalist because, in a sense, you are the expert at your particular publication.

salary: Entry-level salaries begin at around $23,000, while a senior editor makes $28,000 to $45,000 a year.

prospects: The number of magazines dedicated to the environment is growing, but that doesn't mean the jobs are easy to come by. These magazines tend to be in a few concentrated places such as Des Moines or New York. All the major publishers, though, have introduced new titles in this field, and it is possible to get your foot in the door with persistence and a demonstration of genuine interest in the subject.

qualifications: While a journalism degree isn't essential, it is important for an environmental journalist to take a few courses in the basics of journalism and, of course, to have strong writing skills. Many editors of special-interest magazines believe a strong fundamental education in liberal arts is what will serve their readers best. Editors need to have previous experience with special-interest publications.

characteristics: It is important to think critically. Environmental journalists and editors need to be able to translate and present what the experts say for an audience that may have only a vague grasp of the technicalities. You need to be friendly and inclusive when dealing with people, but since many of them (for instance, public relations people) may have an agenda to promote or a product to sell, you have to be firm enough not to let others dictate your job to you. You should also enjoy writing.

Scott Meyer *is senior editor at a national organic gardening magazine.*

How did you get the job?

Scott Meyer has a bachelor's degree in communications. His first job was with a health food trade magazine, and his next was with a weekly newspaper. He became familiar with his current employer through contacts in the business. Scott didn't get the job he applied for, a food editor's position, but was offered a job as a production editor. He later was shifted to equipment editor, which is his current position.

What do you do all day?

Scott has what he calls "flexible" hours, but there's a lot he has to get done. He is, essentially, in charge of articles about gardening tools. He conducts his own product testing and writes several articles and a column for each issue of the magazine, which is published nine times a year. His area of responsibility accounts for between 10 and 20 percent of the magazine's content.

Recently, Scott said he was sending weed whackers to people in warm climates so they could evaluate the tool's performance in time for the spring edition of the magazine. "I'm trying to put together a network of testers around the country,"

HAVE SOME STORY IDEAS READY TO OFFER DURING AN INTERVIEW FOR AN EDITOR'S OR WRITER'S POSITION. IF THIS CAREER IS WHAT YOU REALLY WANT, TRY TO GET PUBLISHED AND ESTABLISH A PORTFOLIO OF CLIPS. SEND IN IDEAS AS A FREELANCE WRITER.

---

he told us, "so I'm on the phone a lot collecting information from them and talking with manufacturers and public relations people about the products. I also spend a lot of time with the art department arranging for what needs to be photographed."

Combine all that with the writing he needs to file, and there are some long days involved, especially as a deadline approaches.

Where do you see this job leading you?

Scott believes he's getting broad experience working in the field of special-interest magazines. He hopes the equipment editor's job at *Organic Gardening* will pave the way for a managing editor or editor-in-chief position at a similar publication. "I want," he says, "a magazine that has some kind of ethic, or perhaps," he adds jokingly, "one that just pays me a lot of money."

71. Editor–Environmental Newsletter

description: The environmental field is honeycombed with specialties. Many of these areas of expertise have their own organizations with their own newsletters and professional journals designed to "get the word out" about the latest developments and advances. An environmental newsletter editor essentially gathers information and keeps people informed. In this job, you are acting as a reporter for a small audience of experts; you'd better know what you're writing about.

salary: Entry-level salaries are $18,000 to $25,000 annually.

prospects: People in this field say they are noticing more journals and publications about environmental matters than ever before. Some are little more than pamphlets, while others run to the glossier, corporate in-house publications. The jobs are out there, but it's a very competitive market.

qualifications: A journalism or related degree is a big help. You should also have a working knowledge of the field your publication addresses. It doesn't hurt to have some course background in biology, chemistry, and the underlying science that affects the environmental field.

characteristics: Resilience and confidence are key. In this job, you have to be willing to research and study someone else's field of expertise. Then you must also have the confidence and resilience to write about it and to edit others who are writing about it. You have to be able to take complicated scientific subjects and distill them into plain English. The experts you interview may be tempted to jump in and manage the journalistic aspects of this job. You have to be comfortable enough in your understanding of the issues to make firm decisions and stick by them.

Jeanne Mettner *is an editor of an environmental digest published in Minnesota.*

How did you get the job?

Jeanne Mettner saw her current job advertised in a local newspaper. With a degree in advertising from the University of Missouri and an interest in public health and medical issues, she thought the job sounded ideal. At school, she specialized in writing and editing health-science articles, and even did a three-month internship at *Health* magazine in San Francisco. She answered the ad, and was hired.

What do you do all day?

Jeanne spends lots of time on the phone, either looking for writers to do articles, talking to writers about a story they're working on, or contacting the 14 members of the publication's editorial board—the people who make sure what she's doing makes sense in scientific terms. (Recently, she was facing a deadline, doing edits on several

> READ ENVIRONMENTAL PUBLICATIONS AND PROPOSE AND SUBMIT YOUR OWN ARTICLE IDEAS. GET PUBLISHED AND COMPILE A PORTFOLIO OF CLIPS.

features and a commentary.) The entire eight-page publication is then sent out to the editorial board members, who comment before Jeanne makes the final edit and mails the issue to the 1,100 subscribers.

Although the publication is targeted for environmental health professionals, Jeanne's editorial direction makes it easily readable without compromising scientific integrity. "We try to maintain language that's not too technical, so that people without a scientific background can read it. Although our audience is very specific, I see myself as communicating to the general public because I know eventually it could end up being on

a desk where anyone can pick it up and read it," she says.

As part of her editorial duties, Jeanne writes "update" briefs and some introductory pieces. She also designs the layout. In between that work, she writes grant proposals, seeking money for the ancillary publication of special issues.

Where do you see this job leading you?

For the immediate future, Jeanne says, she's in a learning mode. She hopes working on this publication will deepen her knowledge of the entire environmental health field. The more she sharpens her expertise, she believes, the brighter her future job prospects become.

72. Editor–Environmental Book Publishing

description: In publishing, an editor prepares and oversees book projects from start to finish. It's the editor's job to conceive ideas, connect with authors, solicit manuscripts, and shape the final text. Environmental issues, developments, and policies are changing daily, and the need to disseminate this information is intense. Depending on the publisher, an editor's responsibilities include acquiring books, negotiating contracts with authors, and many hours of reading solicited and unsolicited manuscripts.

salary: Salaries start at about $18,000 per year. Someone with five years' experience typically makes between $25,000 and $35,000 a year.

prospects: Large publishing companies are expanding to create departments specializing in environmental books. Opportunities also exist with large university presses, which are releasing an increasing number of environmental books and publications.

qualifications: Seeking a journalism or English degree is one way to prepare for a publishing career, but an alternative is seeking a liberal arts degree with a background of writing and editing courses. Obviously, being an editor requires a strong writing background.

characteristics: Aside from good writing skills, a successful editor needs to recognize opportunities and to have the skill to extract information from the right people about the right subjects. A good editor needs to know how to research and gather information and adhere to deadlines. The job often demands tact and negotiation in dealing with authors and answering to an editorial board. A good editor needs to know how to offer constructive criticism.

Heather Boyer *is an editor at an environmental book publisher based in Washington, D.C.*

How did you get the job?

This is Heather Boyer's first publishing job. After graduating with a bachelor's degree in English, Heather enrolled in a six-week publishing program in Denver. That led to an internship with a literary agent in Minneapolis. From there she moved to Washington, D.C., where she saw this job advertised in the newspaper. Heather was hired as an editorial assistant. Those first few months she spent doing administrative work, opening mail, writing letters, and observing editorial meetings. This proved to be a great training ground. Within a year she was promoted to associate editor, and eventually she became an editor.

> ONE SUGGESTION IS TO ATTEND ONE OF THE SPECIALIZED PUBLISHING PROGRAMS OFFERED BY SEVERAL UNIVERSITIES, INCLUDING STANFORD, HARVARD, AND NEW YORK UNIVERSITY.
>
> THESE PROGRAMS ARE USUALLY FOUR- TO SIX-WEEK SUMMER SEMINARS THAT COVER MANY OF THE BASICS OF THE MAGAZINE OR BOOK PUBLISHING BUSINESS. THEY OFFER CONTACTS AND THE MOST CURRENT INFORMATION ABOUT THE PUBLISHING JOB MARKET.

What do you do all day?

One part of Heather's job is to conceive, research, and develop books in the area of community land-use planning and environmental health. Every day she's on the phone with authors, negotiating book contracts, checking on their progress, or prodding them to meet deadlines. Recently, one of her bosses had just come in to talk about final revisions on a book about to be published. After that, Heather wrote a book description for a catalog and later phoned an author to negotiate royalties. It was a full day, as she still had to read a proposal for a book on risk assessment and write a memo about another book slated to be discussed at an upcoming editorial meeting.

Heather's job does require some travel. She attends several conferences a year. At these meetings, she not only learns what the environmental experts are talking about, but also promotes her company's books as she searches for new customers. Back at the office, she follows up on all her contacts from the conference.

All this work doesn't leave Heather a lot of time for uninterrupted reading. She says she finds it easier to read early in the morning or on the weekends at home.

When Heather took this job, she certainly knew about editing and publishing, but she didn't know as much about the specific environmental issues with which she deals now. Heather exemplifies how editors need to be quick learners and adaptable to tackling assignments on unfamiliar ground.

Where do you see this job leading you?

When she started in this business, Heather didn't particularly have a strong passion for the environmental movement, but now she's hooked. She says she's thought of going back to school for environmental health or geography, but she loves publishing and wants to stay where she is, learning as much as she can.

73. Direct Mail Manager— Environmental Book Publishing

description: The direct mail manager's job varies from one publishing firm to another, but overall he or she is responsible for promoting and selling books. Catalogs, which feature listings of the books offered by the firm, are one of the most important selling tools created by direct mail managers. They are an essential marketing tool sent to individual customers, bookstores, retailers, libraries, schools, and universities. Catalogs are usually mailed out several times a year. For smaller specialized publishers, this is their only means to present the product to the customer. It is the direct mail manager's job to decide the best way to feature the books. Direct mail managers also write the catalogs, which include a brief description or synopsis of each book. The manager makes critical decisions about the mailings, such as which customers to target. He or she also monitors response and sales. At larger book publishers, these responsibilities may be divided among several people.

salary: In publishing, entry-level salaries are notoriously low, usually starting at around $15,000. But in the larger publishing firms, a well-experienced direct mail manager who is promoted to positions such as marketing manager or marketing director can make from $40,000 to $70,000 a year.

prospects: As larger publishing companies expand to specialize in environmental books, there is more opportunity in this field. Certainly, environmental organizations use direct mail people for fund-raising efforts.

qualifications: You don't need any one specific degree to become a direct mail manager. People are drawn into this career from journalism, marketing, business, and liberal arts backgrounds.

characteristics: A direct mail person must be self-motivated and detail-oriented. He or she must also be a good writer, and be able to take a large amount of information and distill it into a brief synopsis.

Rebecca Currie *is the direct mail manager for a nonprofit environmental book publisher.*

How did you get the job?

After getting a degree in English, Rebecca Currie enrolled in a four-week publishing program. While there, she saw a job with Princeton University Press posted on a job bulletin board. She interviewed for an editorial position, but she was hired as a direct mail marketing associate. It was a low-paying internship, but it gave her a start and the opportunity to learn the basics. The job involved computer work and assisting designers with layout, proofreading, and tracking sales. When her boss got a call from a small, nonprofit, environmental book publisher searching for people to hire, Rebecca was recommended.

What do you do all day?

"A lot of it depends on what stage of a project I'm in. Two or three times a year I'm in a mad rush to write catalog copy, look over manuscripts, write copy, and get it approved. Authors look at it . . . I rush it to designers, talk to printers, and just get it done," Rebecca says.

The books published by her firm are aimed at professionals in the environmental field, and are to be found in many libraries and general bookstores. Some of this publisher's biggest customers are colleges and universities. Generally, this company's books are sold not as textbooks per se, but rather for supplemental course reading.

"My job is to get word out to people what kind of books we're doing, make people aware of what we're doing, and generate sales," says Rebecca. On a day-to-day basis, her time is spent reviewing manuscripts and writing catalog copy, as well as overseeing design and layout. She's in charge of putting out several catalogs a year for a total of 30 to 40 new books a year. She also arranges the copy and works with the designers who actually do the page layout. Probably the most crucial step is choosing the mailing lists, which are compiled from several sources. In addition to the catalogs, Rebecca sends seasonal announcements, flyers, and brochures promoting books that might appeal to a particular segment of customers. For example, she might create a brochure targeted primarily for university faculty. Part of her job includes tracking and monitoring responses.

Working for a small publisher allows Rebecca freedom to make her own hours and even sometimes work at home. If she's crashing toward a deadline, she'll write at home to avoid the office interruptions and distractions. Although she works with a marketing director, Rebecca is given free rein.

Where do you see this job leading you?

Rebecca says that although she has always been interested in ecological issues, she had never considered an environmental career until working at this particular publishing firm. This job, she says, has definitely raised her level of environmental awareness. That's why she wants to stay where she is. She loves the challenge and opportunity of working at a small publishing firm rather than one of the giants. She's developing new brochures and sees potential for marketing on the Internet.

> **ASK FIRMS OR PUBLISHERS WHERE THEY ADVERTISE JOB OPENINGS.**

74. Environmental Media Associate

description: Understanding environmental issues is one thing; getting the message out is quite another. An environmental media associate helps an organization take the mass media route in trying to raise grassroots support for the group's causes or interests. In some respects, the job is that of a traditional public relations representative, helping the client—his or her bosses—get maximum media exposure for specific events and issues. There is a lot of fielding questions and inquiries from reporters as well as handling the logistics of press conferences and media interviews. But the job is also a strategy position that requires a complete knowledge of the environmental issues at stake and of the political process the interest group is often trying to affect. The environmental media associate is charged with shaping an organization's image.

salary: Starting salaries are around $24,000. A department head, such as a communications director, can earn up to $80,000.

prospects: It's a tough field to get into, but the growth of environmental groups should provide more opportunity. There's always more media work to be done.

qualifications: Generally, people in these positions come from within the organization or have extensive experience in environmental groups. A public relations or communications degree is not mandatory in most cases, since many of those techniques can be picked up on the job. A background in political science is considered a plus. Understanding of the core science beneath the environmental issues at stake, though, is absolutely essential.

characteristics: Good communications skills are essential. You have to be able to communicate the group's message (in a way that will be meaningful to specific audiences) without alienating "outsiders." You also have to be able to prioritize your tasks, and you must have a deep well of self-confidence and not be easily frustrated. Environmental groups are perennial underdogs in the fights they choose. You will be up against smart, well-funded opponents.

Daniel Silverman *is a media associate with a nationally known environmental group based in San Francisco.*

How did you get the job?

Daniel Silverman has a deep passion for environmental issues. While earning a degree in political science, he quit school to work as a campus organizer for one group. He managed to land an internship at the Washington, D.C., offices of a national organization, which is how he made the contacts for his first job as a grassroots organizer. Eventually, Daniel moved to California to direct a coalition of groups that had come together to deal with water issues concerning San Francisco Bay and northern California, building his list of contacts in the environmental movement the entire time. Through these contacts he was able to make the step up into playing a key role with a national organization.

> START WITH THE GRASSROOTS ORGANIZERS IN AN ENVIRONMENTAL GROUP, AND FROM THEM LEARN THE RULES OF THE GAME. TAKE SPECIAL CARE TO OBSERVE HOW THE ORGANIZATION YOU WORK WITH INTERACTS WITH THE MEDIA AND IS ABLE TO COMMUNICATE ITS MESSAGE. NOTE THE DIFFERENCE BETWEEN THE GROUP'S SELF-IMAGE AND THE WAY IT IS PERCEIVED BY THE PUBLIC.

What do you do all day?

A typical day includes an hour of e-mail messaging to various organization offices around the country, a couple of hours on the phone, and at least one conference call—to keep in touch with issues and activities Daniel may need to know about on short notice. There are press releases to write and usually a dozen or so calls from reporters on various issues, looking either for information or for the right person to interview. Recently, Daniel had been overseeing the logistics of an upcoming news conference about environmental civil rights. Was the event listed in all the wire service advisories? Were the speakers lined up? Was the site of the press conference easily accessible to the media? It all had to be handled.

Daniel admits that it's nearly impossible to stay on top of every issue of every regional office of one of the world's largest environmental organizations. He sets priorities and stays focused on the organization's major national campaigns. He's also holding training sessions with their regional offices across the country so that the local activists can better utilize media skills and do their own local public relations work.

Daniel says that the best part of the job is "knowing that you're part of a successful campaign that makes a difference, like when we help get a good environmental law passed. It's gratifying and exciting." On the flip side, he says, the job can become overwhelming in keeping up with all the issues that involve the environment, and in tracking the dozens of environmental bills being considered. "There are so many things happening in the environment that we want to address."

Where do you see this job leading?

Daniel has had several environmental jobs in the past few years, so he says he's happy to enjoy this one and settle in for a while. He believes it's a job he can really develop, and would like to see where this part of environmental work takes him.

75. Public Affairs Officer

(NONPROFIT ORGANIZATION)

description: Depending on the size of the organization, the public affairs/public information officer has a variety of responsibilities. He or she may act as a media liaison, write press releases, arrange interviews, answer the public's questions, notify or alert the media about special events or noteworthy scientific findings, arrange seminars, and give lectures. The public affairs person may be in charge of keeping the organization's name in front of the media or out of the limelight altogether. The job requires writing, be it pamphlets or grant proposals. In some cases the public affairs officer acts as a spokesperson, and is interviewed by newspaper, radio, and television reporters.

salary: Depending on the size of the organization, salaries can start at about $17,000, but for someone with a couple of years of experience, the average salary is in the mid-$20,000s.

prospects: An environmental group can be only as important as its mission and the message it's trying to deliver. Public affairs is an integral department within every environmental group.

qualifications: A degree in journalism or communications with a specialization in public relations is usually recommended. However, several organizations have public affairs people who have risen through the ranks of the group. They've worked as campaigners, grassroots organizers, fund-raisers, and so on. But they all know the organization inside and out and can articulate its mission and message.

characteristics: Public affairs is just that—dealing with the public. Aside from the obvious good communication skills, a public affairs person needs to possess poise and self-confidence. One needs to be articulate and persuasive. Good writing and speaking skills are essential because the public affairs officer acts as spokesperson for the organization. Most of all, one needs to be sincerely interested, even passionate, about the subject one is promoting.

Robert Benson *is a public affairs officer for a nonprofit conservation organization based in Austin, Texas.*

How did you get the job?

"I never dreamed I'd be doing this," Bob Benson is quick to say. This is his first job out of college, and he may have landed one of the most fascinating public relations positions in the country. Bob is the public affairs person for Bat Conservation International. Begun by one of the world's leading authorities on bats, it is the foremost information clearinghouse on this animal, serving as an educational resource about these beneficial but misunderstood creatures. Bob saw the job advertised in the newspaper. He says that although there were applicants more qualified for the job, they weren't willing to promote the subject matter—bats. More importantly, they weren't willing to touch a bat. Yes, his job requires having to handle a tame fruit bat as part of lectures he gives.

What do you do all day?

When Bob was hired, he knew absolutely nothing about bats, so for the first three months he read everything he could about them and sat in on interviews that the executive director of the organization gave to the media. Not only did he learn about bats, he learned what works as a good media sound bite.

As part of his daily routine, Bob arranges interviews between the media and staff bat biologists. He also does many interviews himself. He writes press releases and mails out literature answering the dozens of inquiries a day from the public. He also gives lectures to school and community groups. "I was very nervous, terrified the first time I had to speak before a large group," he says. He quick-

YOU CAN START TOMORROW BY VOLUNTEERING WITH A NONPROFIT ORGANIZATION. ASK TO HELP IN THE PUBLIC AFFAIRS OFFICE. YOU MAY BE ONLY LICKING ENVELOPES, BUT YOU'RE ALSO MAKING CONTACTS AND WATCHING FIRSTHAND HOW THE OFFICE WORKS.

ly got past that, as he's given some 100 lectures in the two years he's been with BCI.

The advantage to working with a relatively small nonprofit organization—BCI has only 24 staff members—is that Bob's called upon to pitch in on whatever needs to be done. This year he traveled to Pennsylvania to assist one of BCI's staff biologists in conducting a workshop. At night, they would catch bats and teach other wildlife biolo-

gists how to identify bats by species and sex. Summer nights find him under the Congress Avenue Bridge in Austin. It's become quite a tourist site because up to 1.5 million bats collect there each night, during the peak of the season. Bob gives talks and answers questions.

Where do you see this job leading you?

Bob hopes to advance into a managerial position with a larger organization, but doubts he'll ever do public relations for a more fascinating and unusual subject.

76. Environmental Instructor

(PUBLIC AQUARIUM)

description: An environmental instructor uses the database and physical plant of a museum, aquarium, or nature theme park to teach lessons about the underlying science displayed by the facility. In this job you have to be able to work and communicate with scientific specialists and the public. There is usually a strong element of showmanship connected with this job that transcends the role played by a simple nature interpreter. The instructor may be required to construct and deliver these education messages off site for educational and/or marketing purposes.

salary: Starting salaries vary depending on the aquarium. Some aquariums' starting salaries are in the low $20,000s per year, but experienced applicants may command up to $30,000 annually.

prospects: The opportunities are somewhat limited, but once you get into the field there's a good internal market for people who exhibit talent. There are roughly 90 aquariums and commercial water parks operating in the United States and several new ones in the planning stages. There is turnover in the field, so keep trying. Natural history museums and wildlife parks are also sources of this type of employment.

qualifications: While no specific degree is generally required, you will need a working knowledge of the field's underlying science, as well as some experience or course work in education. Familiarity with theatrical arts and techniques is also helpful, although many facilities offer this training to the people they hire. You should be a certified scuba diver.

characteristics: This job requires initiative and a tremendous amount of flexibility. You will be doing everything from building and delivering educational programs, to training employees. In some instances you will be taking your facility's message on the road in an effort to generate interest in the research or exhibits your employers have developed.

Scott Stratton *is an environmental instructor at Monterey Bay Aquarium in California.*

How did you get the job?

Scott Stratton wasn't sure what he wanted to do. His first love was the ocean. He had a bachelor's degree in biology and had discovered a talent for teaching during a winter spent instructing disabled people how to snow-ski. As he worked his way up the West Coast doing job interviews, a potential employer told him about the opening at the aquarium where he now works. That was three years ago. He's grown in the job and been given new responsibilities.

What do you do all day?

Scott says he doesn't have a typical day. Because he takes marine animals such as sea stars, urchins, and hermit crabs on the road for his presentations, he sometimes gets up before dawn to help load the truck used to transport the creatures and the other visual aids. Recently, the life support sys-

> **BE WILLING TO MOVE TO ANOTHER GEOGRAPHICAL AREA. YOU HAVE TO BE WILLING TO GO WHERE THE JOBS ARE.**

tems in the truck had given out and the animals had to be transferred back into tanks at the aquarium on short notice. Scott visits lots of schools and often meets school field trips at the aquarium to guide them through the exhibits. He says he's written elaborate 20-minute theater pieces using music and characters to illustrate some of the aquarium's lessons. In the summer months he visits migrant farm camps in the state's interior, and he has even gone into prisons and juvenile detention facilities. His duties include training and co-ordinating volunteers, and collecting and feeding the animals for the exhibits. Some days are spent entirely in the water wearing scuba gear.

Where do you see this job leading you?

Scott admits it's tough to move up in this field unless someone dies or decides to quit. He says he's probably poised to move into a coordinator's position somewhere else, although he still dreams of being a marine resources manager working for the state or federal government.

77. Environmental Educator

description: An environmental educator, simply put, is a teacher. But it's a career that encompasses a wide range of opportunities. Environmental educators work as teachers and professors in traditional classrooms. Others work in outdoor education as naturalists and guides at nature centers, zoos, aquariums, summer education programs, and youth camps. Environmental educators also work as public information specialists for government agencies and environmental advocacy organizations.

salary: Teachers are notoriously low-paid, and there is no exception in the area of environmental education. Salaries are not as high as in other environmental careers. They can start as low as $12,000 a year and can rise to the low $30,000 range in the nonprofit sector, while administrators in the private sector can expect salaries of around $40,000. College and university professors' salaries can range anywhere from $40,000 to $75,000.

prospects: Environmental education is a rapidly growing field. There is a large demand for these teachers, as environmental classes are more popular than ever. Several states have recently approved or are in the midst of considering environmental education laws that require schools to teach students about environmental issues. In terms of growth, it's an area worth considering. However, it takes an entrepreneurial spirit to convince budget-strapped school districts to seriously pursue these courses as part of the curriculum. There's a growing emphasis on urban environmental education.

qualifications: A bachelor's degree in education, environmental education, or natural science is usually required. However, most employers prefer that the degree be in one of the natural sciences rather than in education. Teaching experience or education study is still a prerequisite for the good jobs. A teaching certificate is usually required at elementary and secondary schools.

characteristics: Environmental educators need to be enthusiastic, creative, and dedicated professionals. Critical thinking skills are also a plus.

Jeff Cole *is an environmental educator in New York City.*

How did you get the job?

You probably wouldn't expect to find a high school dedicated to and specializing in environmental studies in the heart of the concrete jungle that is New York City, but there is one. The High School for Environmental Studies serves as a magnet school that offers environmentally related courses not available at other schools. It's here that Jeff Cole has carved out a unique teaching role for himself. He is the director of a student internship program that will someday serve as a model for schools throughout the country. Jeff's job is to coordinate and monitor some 140 students in environmentally related internships throughout the school year and summer.

> **VOLUNTEER OR TRY TO GET AN INTERNSHIP WITH A NATURE CENTER, MUSEUM, OR ZOO.**

Jeff didn't plan to be a teacher. In fact, he says, he never dreamed he'd be doing this. With a master's degree in environmental economics, he planned to work for a Washington-based organization or overseas in international business. As a Fulbright scholar, he worked in Poland in the area of economic reform.

Upon returning to the United States, Jeff got a job with the National Environmental Education and Training Center, where he heard about Teach for America, a national program that places people with degrees in fields other than education in teaching jobs. Before he knew it, Jeff was teaching social studies to 10th-graders and coordinating the internship program on a part-time basis. He seized a great opportunity to expand the internship program and make it his full-time job. A special school fund-raising program obtained the money to pay his salary.

What do you do all day?

Jeff spent the first few months on the job working the phones and meeting with clients or potential collaborators, trying to find new and more internships for the students. Since the internship is a volunteer program, he puts the word out to students, recruiting participation and matching the right students with the proper work. A typical day begins with Jeff making phone calls, checking on new internship possibilities and on the students themselves—which interns didn't show up for work or what problems others are having. He monitors their progress and, with the help of other teachers, meets with the interns once a week in advisory sessions. He acts as their mentor, answers questions, and helps them write resumes.

Jeff's taking the job even further. He's now developing an environmental career center within the school. He's begun by purchasing books to stock the center's library. The center's purpose is to help students identify and choose environmental careers and prepare them to enter the job market. At the center, they will learn resume-writing and interview skills. About the job overall, Jeff says, "It's unbelievably rewarding. That was the motivation for me getting into it and I think my expectations are more than fulfilled. It's been a wonderful, wonderful job."

Where do you see this job leading you?

"The model program that we've created here could be applicable in a wide number of areas, and I wouldn't mind taking the show on the road and continue consulting for this kind of setup in different areas," says Jeff.

78. Independent Documentary Filmmaker

description: The independent environmental filmmaker is one who wears many hats. He or she is often a producer, director, writer, photographer, and sometimes even editor and promoter, all wrapped in one busy person.

The process of making a film starts with an idea. Then comes the hard part: finding the money to make the film. This is often the most grueling and discouraging part of the job. It's true that talent and an undefinable "eye" for filmmaking are crucial, but there are many other qualities needed to create a successful nature film.

To attain at least a minimal level of success, the film must be marketable. Often the most laborious work is the preproduction and postproduction phases, rather than the shoot itself. An environmental documentary filmmaker spends months in preproduction—formalizing a script and arranging shooting schedules, equipment, and locations. He or she must hire video and audio crews. Afterward, there's the editing of pictures, sound, music, and narration.

As difficult as it is to get project funding, it's even tougher to find a broadcast outlet. The market for nature documentaries is shrinking as fewer television networks and local channels choose to broadcast films with environmental subjects.

salary: Not many independent filmmakers can make a living creating only environmental films or documentaries. They usually supplement their income by tackling other commercial projects or even by working for others in the film industry as freelancers.

prospects: Sadly, at a time when there are more independent filmmakers interested in this subject, the industry is changing. A trend toward television media conglomerates has left these producers with even fewer broadcast outlets.

qualifications: Extensive training in several media disciplines is required. Television and film courses are obvious ones to take, but you should also be able to write.

characteristics: Being an independent environmental filmmaker takes an extremely committed, driven, and relentless personality as well as someone talented in film craft. One documentary producer advises that you also need to "demonstrate hustle and thrive on rejection." It also takes patience, as a film can take years to make.

Frank Green *is an independent environmental documentary film producer.*

How did you get the job?

Frank Green has an extensive and distinguished filmmaking career. He has worked as a journalist, cameraman, and field producer for NBC News, CBS News, and the BBC. As an independent filmmaker, he's lived in Hong Kong, where he produced and shot promotional, industrial, and news films. His clients include the BBC, CBS News, and the Canadian government. He photographed and co-produced *Tongpan,* a Thai feature film about a Thai farmer, which the British Film Institute called "one of the most important 'Third World' films of the decade." He won critical acclaim for his film *The Forest Through the Trees,* an independent television documentary about the crisis in California's redwood forests. It's been televised on PBS stations and the Discovery Channel and is now in nationwide distribution.

What do you do all day?

A typical day is spent either in his office, which is in his home, or on the location of a shoot. When Frank's in the office, a lot of his time is spent on the phone or on the computer, taking care of the many administrative details of his small independent company. Most of his time, he says, is dedicated to project funding—that is, seeking people to "back"—donate money for—his film projects. He's also keeping established backers abreast of the film's status.

Although Frank acts as the creator and executive producer, film projects involve many professionals. He hires a writer, an editor, a soundperson, and others.

> **GET AN INTERNSHIP WITH A PRODUCTION HOUSE OR AN INDEPENDENT FILMMAKER. BE WILLING TO WORK AS A GO-FER, GETTING COFFEE OR DRAGGING CABLES.**

Recently, Frank was just finishing editing a one-hour film about a timber conflict in the Sierra Nevada. Now he faces the arduous work of finding an outlet to telecast his film. "The outlets for this kind of show are few. There's Turner, Discovery Channel, and to some extent there's PBS. I may have to approach every PBS station individually, which is an enormous amount of legwork. There are a few other outlets, but that's pretty much it for national outlets," he says.

Without a doubt, finding the money to make these films is the most exasperating part of the business. Rarely will an independent documentary producer get all the money up front. Instead, it is often a matter of nickel-and-diming his or her way through a project.

Where do you see this job leading you?

Frank poured an enormous amount of time, work, and dedication into his last project, *The Forest Through the Trees.* Despite its critical acclaim, in terms of money, the film just about broke even. So now Frank's rethinking the types of films he's making. In the past he's produced issue-oriented environmental documentaries. Now he wonders if shorter-length (less than one hour) films would be more marketable. "The kinds of films that do better are the kinds of things that *National Geographic* makes . . . . with sharp teeth. Sharks are a winner."

79. Nature Photojournalist

description: A nature photojournalist takes the skills of a photographer into the wilderness in order to document life there or to illustrate research being conducted on site. This is a specialized skill useful to only a handful of magazines, newspapers, and professional journals with a strong commitment to environmental reporting. The work is active and sometimes exciting, but is almost always the province of people who already have established skills in journalism and photography.

salary: Salaries aren't much when you're getting started. However, eventually the sky's the limit, especially for photographers able to generate good volumes of salable shots that can be marketed through commercial stock services.

prospects: Slim. At a level where it's profitable, only a relative few work for magazines such as *National Geographic*. Even with knowledge, hard work, and aptitude, there is an indefinable quality that generates success in this field.

qualifications: To enter this line of work, you should have experience in photojournalism or, at minimum, a working knowledge of photographic principles and equipment. You should also have an understanding of environmental or biological science and familiarity with outdoor living, including basic deep-woods survival and scuba diving.

characteristics: Successful people in this field say patience, physical endurance, and instinct are all essential at various stages of the job. The patience first comes into play during research and planning sessions that may involve learning about a totally new subject in enough detail to map out how you'll want to photograph it. There are also extensive logistical arrangements to be made, so you have to have a head for that type of thinking as well. Physical endurance becomes essential when you have to accompany researchers or naturalists into the habitat. That could involve extensive hiking, rock climbing, or a lengthy sea voyage. As you live with your "guide," you will again need patience as you wait for the moment when his or her work will present a photographic opportunity. Instinct, of course, is the imponderable quality that will tell you it's time to take a picture. Sometimes, the success of a monthlong trip might be measured by the quality of a shot taken in a single split-second opportunity. You might find yourself working in camps with less than ideal or cramped facilities, so you need to be able to respect and get along with others.

Flip Nicklin *is a nature photojournalist working primarily for* National Geographic *magazine, based in Washington, D.C.*

How did you get the job?

Flip Nicklin always wanted to be a scientist, but teachers and guidance counselors kept telling him he didn't have the math and science proficiency. His father was a cinematographer, so Flip was exposed to the basics of that business early on. But that kind of work never appealed to him either.

In his early 20s, Flip was working as a dive instructor in San Diego and found himself teaching a group of *National Geographic* photographers and writers how to scuba dive. They asked him to come along on a trip to the refuge islands of Hawaii. He happened to bring along a camera, took several pictures, and actually got them published in the magazine. The first photograph he got published in *National Geographic* was an unusual shot of an albatross jumping over a seal.

Flip never had any formal training in photography, and says he wishes he'd taken journalism courses in school. Because of the lack of training

> SIMPLY BEGIN. IF YOU HAVE AN INTEREST IN PHOTOGRAPHY, BUY, BORROW, OR RENT A CAMERA AND START SHOOTING. COURSES IN PHOTOJOURNALISM AND ENVIRONMENTAL SCIENCE MAY INTRODUCE YOU TO THE NETWORKING POSSIBILITIES THAT CAN LEAD TO INTERNSHIPS OR SUMMER JOB OPPORTUNITIES. YOU SHOULD SEEK OUT A MENTOR—SOMEONE ALREADY DOING THE JOB, WHO CAN TEACH YOU WHAT HE OR SHE KNOWS AND HELP YOU FIND OUT IF YOU HAVE THE INNATE TALENT TO TAKE A SERIOUS STAB AT THIS COMPETITIVE FIELD.

in journalism, he says, his start in the business was "a little rough." After getting that first photograph in *National Geographic,* he started shooting pictures for aquariums and football games for the San Diego Chargers. It took three years after getting that first photograph published before he began to get regular assignments from *National Geographic*. His first assignment for them was shooting whales and sharks. He recently received acclaim for a rare photograph of a white sperm whale taken in the Atlantic Ocean off the Azores.

"It was like finding Moby Dick," he says.

What do you do all day?

The simple answer is "Wait and prepare." Flip says the preparation for "a shoot," as it's called in the business, can take months. He says one recent job in Central America involved a period of 54 days without taking a single shot.

Flip spends a minimum of four to five months in the field each year. He recently moved to Washington, D.C., from the

Puget Sound area to be closer to the magazine's headquarters and to cut down on the amount of time he was spending on commercial jets.

There are a million things to do, Flip says, but the main task he has at every moment is to be ready when luck presents an opportunity. His famous shot of the white sperm whale occurred at the end of a long trip to the Azores, on one of the final days out. The whale, he says, was in his sight for no more than four seconds. On a recent day, he was working at the office, planning the details of assignments that would soon begin. He says they will take him to both the tropics and the Arctic.

Where do you see this job leading you?

At the age of 47, Flip says he knows he's not as quick as he used to be and is preparing to move into other areas of photography. There may be seminars for him to teach. He is about to begin promotional work on a book of marine mammal photos he's taken during his career.

80. Video Lab Technician

description: This is a very new and exciting job in the environmental career job market. It's as current as the cutting edge of technology. This new job combines science, computers, and video technology. Scientific observations are no longer solely recorded in the form of written data. They are also recorded on film and videotape. Someone has to interpret, categorize, and log countless hours of videotape. It's not merely a job of screening tapes; it calls for the scientific expertise to extract, categorize, and cross-reference data for several scientists. Essentially, you'll watch endless hours of some of the most fascinating or dull videotape. The job may require you to enter data on the computer and catalog it. Someone hired for this job might be required to create computer cataloging systems and methods.

salary: With a bachelor's degree, entry-level salaries begin at $18,000, and with a master's, at $23,000.

prospects: Video technology is one of the fastest growing and most innovative career fields on the job market. It's so new that some research projects might not even know they need someone in this position until you suggest and create it. Research institutes and universities may hire these professionals to work on various projects that require extensive videotaping or filming.

qualifications: If you want to work on ecological or environmental projects, a good science background will set you apart from those with only video or film school experience. Those without a scientific background won't know what to look for when screening tapes. A knowledge of computers helps, as you may be required to enter scientific information into a database.

characteristics: It takes a detail-oriented and very organized person to do this job well. It also requires someone with the ability to look at video analytically, to interpret the images, and to translate them into useful information for the scientists on the research team.

Dr. Judith Connor *is a video lab director and researcher of deep-sea exploration at the Monterey Bay Aquarium Research Institute.*

How did you get the job?

With degrees in botany and chemistry, Judith Connor first landed a dream job of sailing beautiful Caribbean waters studying coral reefs. On the three-year expedition, she worked as a field technician for a researcher with the Smithsonian Institution. She quickly became a leading expert on Caribbean algae.

Then Judith decided that instead of being someone else's technician, she wanted to head up her own projects. So she went back to school and earned doctorates in botany and ecology. She did her postdoctorate work at the Stanford University marine lab. That led to a full-time job at the Monterey Bay Aquarium, next door. At the aquarium she wrote exhibit texts and two natural history books.

Judith eventually made the leap to a job as a research associate at the research institute. Even though she wasn't an expert on video, she applied for the video lab director position. She was confident that her scientific background would get her over any hurdles. She jumped into the job at full speed.

What do you do all day?

Much of her day is spent watching either a video television screen or a computer monitor. The institute uses a remotely operated vehicle (an unmanned submarine) to explore the deep sea. The submarine, filled with scientific tracking devices and cameras, transmits live pictures from 6,000 to 12,000 feet down.

"We see things no one has ever seen before," Judith says. The show she watches is an underwater marvel filled with color and drama unlike any movie or television show. "If you could just see some of these images. There's science and there's art, and this is both."

She actually watches about four hours of videotape a day. She collaborates with other scientists to determine new species, new behaviors, and all other pertinent data. Currently, she's working with a computer software engineer to develop a system for better access to all the information on the videotapes.

Does watching the videos ever get boring? You can fast-forward through the boring parts. But more often than not, the videos are simply breathtaking. "We still 'ooh' and 'aah,'

DON'T JUST RELY ON COLLEGE JOB PLACEMENT OFFICES. CALL ALL YOUR PROFESSORS AND ASK THEM FOR HELP AND DIRECTION. THEY MAY HELP YOU GET YOUR FOOT IN THE DOOR FOR THAT FIRST JOB.

and whenever something really fascinating pops up on the screen, we all drop what we're doing and race to the monitors to see."

Where do you see this job leading you?

Judith wants to stay right where she is. She balances full-time work with being the mother of a teenage son. The scope of her work promises to widen, as the institute hopes to promote collaborations with other research and educational institutions. They want to take their live pictures and data onto the Internet. "Soon people at home with their computers will be able to see this video and access data and be scientists like we are."

Judith will be adding a new project to her roster when she begins teaching a marine botany course at Stanford University.

81. Environmental Activist

description: They are the ones on the front lines trying to improve, restore, and protect the environment.

One day the activist may be merely answering the phones at the local office, or he or she may be speaking on an issue at a community board meeting or testifying before Congress. It's the activist's job to arrange events such as speeches and seminars or the more headline-grabbing events such as protests, demonstrations, or boycotts. Almost every activist, or "campaigner," as they're called in some organizations, starts as a volunteer. These aren't 9-to-5 jobs. You might find yourself working very late coordinating volunteers, and confirming last-minute details for an early morning protest. Count on working weekends, too.

salary: As an environmental activist, you can expect a starting salary of $15,000 to $25,000 per year.

prospects: Staff positions with environmental organizations don't come up often. Such groups don't usually have many people on staff and most of these organizations are short of money. As grant money proves to be difficult to get, groups are consolidating jobs and relying more heavily on volunteers. But don't get discouraged; a truly creative and dedicated person will break through.

qualifications: No specific types of degrees are required for this type of work, but backgrounds in the sciences, business, or marketing are helpful. As an activist, your most important qualification is being committed to your cause.

characteristics: To win one of those rare staff positions, one must be smart, dedicated, creative, aggressive, and tremendously energetic.

Mark Floegel *is an activist who works for Greenpeace.*

How did you get the job?

Mark Floegel started as a newspaper reporter, but decided he wanted to work in more of an advocacy-type role. His first job was canvassing door to door for Ralph Nader's Public Interest Research Group. "Even though I had a career and education, I was willing to learn the hard way by knocking on doors and talking about issues. It's a very real way of paying your dues."

The most important thing about being an environmentalist is dedication and selflessness. Mark made barely enough money to pay his bills, but his hard work didn't go unnoticed. He was hired on staff as a canvass director, and then moved into public affairs. This eventually led to his getting hired at Greenpeace.

What do you do all day?

Mention Greenpeace and most people automatically think of the daring activists in boats on the open sea desperately trying to stop whaling or to block French nuclear testing. But most of their workers never get on boats, and the issues they tackle are wide-ranging. Greenpeace is fueled by aggressive, media-savvy campaigners who are well versed in delivering their message and eliciting change, often through direct action.

Mark works on Greenpeace's chlorine campaign. His mission is to convince paper mills to make paper without the use of chlorine, which otherwise results in a discharge of dioxins into water. It's his job to try to come up with ways to force the paper mills to change. After exhaustive yet unsuccessful discussions with the paper mills'

> SIMPLY PUT, VOLUNTEER. MOST NONPROFIT ORGANIZATIONS LOOK TO STANDOUTS IN THE RANKS OF THEIR VOLUNTEER STAFF WHEN FILLING PAID STAFF POSITIONS.

owners, Mark says, it was time for more aggressive action; he arranged events such as plugging up paper mills' discharge pipes.

For Mark there is no such thing as a typical day. He says he does a lot of writing and reading. Recently, he'd just spoken to a grassroots activist in Oregon regarding an issue there. That was followed by conversations with a scientist looking for a copy of one of Greenpeace's reports; the environmental director of the Oneida Indian nations, concerned about discharge from a paper mill; and environmental consultants and investors wondering

what Greenpeace's position is regarding particular paper mills. Floegel recently spearheaded the move to print Greenpeace's version of *Time* magazine on environmentally sound chlorine-free paper to prove to *Time* that it can be done. On the grocery store shelves you may have noticed more unbleached products such as unbleached paper coffee filters. That's directly due to Mark's work.

Where do you see this job leading you?

Mark can't imagine himself being anything but an activist. But as he's beginning to see progress with the chlorine issue, he may soon switch to heading another campaign within Greenpeace. As a trained journalist, he says, he'd like to start writing again, but this time it would be advocacy writing.

82. Grassroots Organizer

description: Grassroots organizers help citizens groups unite around particular issues. The grassroots organizer helps the group formulate its mission and develop an effective strategy toward reaching the goal. The goal may be to prevent change or to initiate it. The grassroots movement is the basis of political and environmental change in this country. Most organizers work on a volunteer basis, but there are full-time salaried grassroots organizers working at many of the national organizations. These organizers are usually assigned a region where they help decide the issues to be tackled and then develop strategies. They help organize local groups from already established membership rosters, or they begin campaigns to recruit new members. Experienced grassroots organizers help groups define issues, develop campaign plans, and implement those plans.

salary: "Salary should never be the motivating reason you get these jobs," says one grassroots organizer. Salaries depend on the organization. Starting salaries at national organizations, depending on their membership and financial strength, are around $20,000.

prospects: Prospects can be very tight. One paid activist tells us that there were 140 applicants for her job. Her organization currently has a job opening, for which she anticipates at least 200 applications.

qualifications: Nearly all full-time paid grassroots organizers started their careers by volunteering with organizations. This profession requires a personal calling to activist means of instigating change or protecting the environment. You will also need to be able to write concisely, clearly, and creatively. The job often requires you to read and digest complicated technical documents and to translate that information to members.

characteristics: A successful grassroots organizer needs to be nurturing and patient. The job requires a constant distilling of information and differing opinions. The two most important qualities that a grassroots organizer must have are to be a good listener and to be an intelligent speaker.

Betsy Buffington *is a grassroots organizer for a national environmental organization.*

How did you get the job?

Betsy Buffington started by volunteering with campus organizations while she was in college. That's where she realized that she had the "people" skills and the creative skills to choose this line of work as a career. After getting a degree in Latin American history, she began working at a copy-printing store. One of their clients was the Wilderness Society. Through the grapevine, Betsy heard about a clerical job there and was hired. After a couple of years, she was laid off and made the jump to the national conservation group for which she now works. She didn't have much experience for that first job, but, she says, her enthusiasm and ideas got her in.

What do you do all day?

As a regional director, Betsy works with activists in six states. She follows priorities or issues directed by the national office as well as any cries for help from local community members. Her region is large and rural, so her job does require traveling. She often travels to small towns to organize meetings and to teach members how to tackle issues of their concern. "Sometimes you run into people who have experience as staff for another organization, but more often than not they are novices when it comes to conservation efforts," says Betsy. At those meetings, she'll put the ideas on the blackboard, then go through each one, offering advice on how to develop a campaign strategy and resources. "Fundamentally, we play the role of organizers and offer guidance," Betsy says.

On any one day, Betsy may get a phone call from a local chapter concerned about a possible threat to a national forest, and then a call from a group concerned about a wolf recovery plan in Yellowstone National Park. Much of Betsy's time is spent reading technical and scientific reports and studies. She then has to translate the complicated information into concise and understandable reports for the membership.

> **GET INVOLVED WITH AN ISSUE AT THE LOCAL LEVEL. IF YOU'RE STILL IN COLLEGE, CHECK OUT CAMPUS ORGANIZATIONS.**

"I think my favorite part of it is the actual grassroots organizing, ... going out and meeting with people ... helping them to develop their skills to address the issues that really concern them. There's a real sense of empowerment on their part," she says. The downside of the job is sometimes dealing with the personality clashes between group members; some people become very passionate about certain issues and causes. Betsy advises that it takes an incredibly flexible person to deal with all the very different and sometimes difficult personalities.

Where do you see this job leading you?

Betsy has been in this job four years and wants to stay where she is. She loves the job and that it allows her to live in a beautiful part of the country.

83. Lobbyist

description: The single mission of a lobbyist is to get legislators on the side of the cause or issue he or she represents. Lobbyists try to influence and bring pressure on lawmakers to promote and, more importantly, vote favorably on certain policies. They do this by meeting, interviewing, and cajoling. Lobbyists seek interviews, write letters, direct phone campaigns, and, if necessary, bring other pressures to bear on legislators. The image of lobbyists is that they're smooth talkers who never back down from their political stance, and in a limited sense, that's true. But they must be well researched and well versed in the issues and causes they represent. Often they may catch only a few precious minutes with a legislator, and that's all the time they have to present their side. Many lobbyists view their job as educators about a cause or movement. Depending on the organization, the lobbyist's responsibilities might also include writing reports, articles, and press releases, and coordinating press conferences.

salary: Lobbyists for nonprofit environmental groups, as might be expected, get paid less than those working for many of the well-funded private industry groups. In the nonprofit world, lobbyists start in the middle $20,000s and with experience reach into the high $40,000s.

prospects: Prospects usually depend on the fiscal health of the environmental groups. The current congressional threat to the environment has sparked a backlash and has increased membership in many environmental groups. When membership is up, people are contributing more money; thus, the organizations can afford to hire more people, including more lobbyists.

qualifications: Lobbyists come from all walks of life and education. Many are lawyers, but that's not essential. A political science or law background can be helpful for maneuvering your way through the system, however.

characteristics: This is not a job for the shy. It takes an aggressive, persuasive, disciplined, sharp, well-spoken person to be an effective lobbyist. One should also be decisive, yet flexible.

Anna Aurilio *is an environmental lobbyist for the U.S. Public Interest Research Group (USPIRG) in Washington, D.C.*

How did you get the job?

Anna Aurilio didn't just walk in the door and say she wanted to be a lobbyist. She graduated with a physics degree in 1986 at the height of the "Star Wars" weapons defense program, but, she says, "I knew I did not want to go build missiles...those were the jobs that were open."

Unlike many of her colleagues, she decided against graduate school or working for an environmental consulting firm, so she answered a newspaper ad for a staff scientist job at Massachusetts PIRG. PIRGs are nonprofit, nonpartisan, nationwide consumer and environmental watchdog groups. They embody grassroots organization and citizen empowerment.

The staff scientist job wasn't an easy job to get. "I was just out of college with no experience and they wanted someone with experience. I talked my way into that job," Anna says.

Little did she know that she was already exemplifying the persistence, energy, and convincing style of a lobbyist. As a staff scientist, she investigated industrial and municipal discharges, examining a wide range of industries, from sewage treatment plants to paper mills to steel mills. Working during the day and going to school at night, Anna earned a master's degree in environmental engineering. Through her connections, she heard about the lobbyist position and gave it a go.

What do you do all day?

You have to run fast to keep up with Anna as she darts around Capitol Hill, running from meeting to meeting, talking with politicians and legislative aides. She starts the day stopping by the office to revise a press release, prints it up, makes copies, and packs them into her briefcase. She dashes to a hearing of the Senate Natural Resources Committee, but at the door is told she can't get in because it's full. That doesn't stop her, as she manages to talk her way in. During the noon break, she talks with reporters, distributes press releases, and speaks with legislative staff people to see which amendments are scheduled for votes. Then it's back to the office to work on

> **START AS A CANVASSER FOR AN ORGANIZATION. THIS IS A WAY TO PICK UP THE BASIC PERSUASIVE SKILLS REQUIRED TO BECOME A LOBBYIST. IT'S A GREAT WAY TO TEST THE WATERS FOR THIS CAREER.**

a new project, research current projects, and return phone calls. Somewhere in between, she fits in three meetings with Senate staffers to check their positions on other bills.

In describing her job, Anna says, "As a lobbyist, I'm harnessing the power of the grassroots to exert pressure on the political process. It's awesome and really exciting." Being a lobbyist can seem to be an intimidating job, but Anna reminds nervous interns that when she first started, she didn't know more than five senators' names. She's come a long way, as people now come to her for advice on strategy and tactics. Recently, Anna was disheartened by the fact that a national environmental protection bill had been defeated only hours before. It's part of the highs and lows of this fervent career.

Where do you see this job leading you?

Anna sees herself staying in the public interest sector, but says she'd like to move into work that combines her political savvy with her scientific knowledge. Her hope is that more technically trained people will consider jobs in the public interest.

177

84. Canvasser

description: A canvasser is the door-to-door salesperson of a movement or organization. Canvassers are employed in all areas of political action, but in the environmental field they are key fund-raisers as well.

The job involves going out into neighborhoods to interest people in environmental issues, convince them to join various organizations, and cajole them to donate money to the cause. This is an excellent way to get into the field of environmental action. It can be your foot in the door into this arena. This entry-level position is not for the meek, but because it involves raising money, it can lead to bigger things.

salary: Base rate in a large city would be $200 a week, plus a percentage of the money raised over a minimal target amount.

prospects: Excellent. Heavy turnover means there are almost always openings.

qualifications: No formal qualifications. Training is always offered to new applicants, but verbal skills and an ability to grasp the issues are essential. You must be personable, assertive without being too aggressive, and resilient enough to bounce back on a day when everyone slams the door in your face.

characteristics: Friendly, outgoing people have an edge in this line of work. You need to be articulate and to have good communication skills in order to discuss the issues with the people you will meet. It helps to be committed to the environment. Since you will meet a number of rude individuals in this job, it helps to have a thick skin as well.

Kirsten Williams *is the canvass director for a public interest research group based in Washington, D.C.*

How did you get the job?

Kirsten Williams answered an ad in the paper. She was waiting tables at the time, and the idea of getting back in touch with her concern for the environment was appealing. Kirsten says she'd been interested in environmental causes before she started college, where she had been working toward a degree in psychology, but had to quit because of financial pressures. She found the job listed under the heading "Activism."

What do you do all day?

This is mostly night work, approaching people in their homes in the late afternoon and evening. The canvassers come to work at two in the afternoon, but Kirsten's day starts earlier since she's a supervisor and needs to handle paperwork and scheduling and interview new job applicants. "We knock on at least 75 doors every night," Kirsten explains. "Out of the 75, you usually talk to 35 people, and the aim is to get five

> CHECK THE NEWSPAPER WANT ADS. YOU CAN ALSO DIRECTLY APPROACH THE APPROPRIATE ORGANIZATIONS IN YOUR REGION (GREENPEACE, LOCAL PUBLIC INTEREST RESEARCH GROUPS, THE SIERRA CLUB, ETC.).

to 10 involved." They have minimum nightly dollar goals. The canvassers work to get people involved at a deeper level than just a financial contribution. They "pitch" information regarding specific issues, such as the group's campaign to prevent oil drilling in the Arctic National Wildlife Refuge.

Canvassers are assigned specific neighborhood blocks. Before heading into the field, they practice their skills through role-playing exercises. They rehearse their pitch and receive the latest information on all the issues they must represent. Canvassers are expected to be well versed in the issues and to speak sincerely about them. "We tell them [potential members] the issues that we're working on, why we're working on them, and why it's urgent that they get involved," says Kirsten.

The canvassers are the backbone of an organization, especially since tackling many issues is an uphill battle. "They [the members] are what give us strength to win these campaigns. Our opponents are strong; for every advocate we have, they have six lobbyists working against us. We're being outspent 1,000 to one. Some people see it as just a job. But what has gotten a lot of people involved and committed is the cause," Kirsten says.

Where do you see this job leading you?

Kirsten hopes to eventually run her own office, perhaps out West. Good canvassers are usually promoted quickly to field supervisors, and then on up the line. Kirsten says she got her first promotion within a month of being hired. She says a lot of middle-level people working in environmental organizations rose through the ranks from entry-level jobs as canvassers.

85. Environmental Victims Advocate

description: A neighborhood suspects something toxic is in their water. A chemical plant plans expansion in a community. Who can community citizens turn to for help? One option, aside from local environmental groups or lawyers, is someone called an environmental victims advocate. It's his or her job to arrange or conduct sampling and to investigate the nature of the possible contamination and its effects on a community. The environmental victims advocate is familiar with not only the testing processes but the legal process as well. He or she must be familiar with various agencies' methods, rules, and regulations.

salary: You're not going to get rich working in the nonprofit sector, but as an independent consultant you can earn from $30,000 to $60,000 per year.

prospects: There aren't many people doing this kind of work. That's not to say it isn't needed, but you'll probably have to carve your own niche.

qualifications: A background in one of the sciences is recommended. Experience with regulatory agencies is also helpful.

characteristics: Perhaps the most important characteristic needed to help these kinds of clients is patience. You need the patience to deal with diverse groups—people who are often scared, anxious, and angry. You also need perseverance; you can't take "no" for an answer. There might be times when a state regulatory agency slams the door in your face and refuses to give you requested information. That's when you have to know how to negotiate your way through the bureaucracy and get what you need.

Wilma Subra *is an environmental victims advocate.*

How did you get the job?

Wilma Subra, who has master's degrees in microbiology and chemistry, worked at a research institute for 14 years before deciding to open her own, unique consulting business. At the research institute, Wilma was in charge of the environmental sciences analytic department. She and her team conducted and tested water and soil samples. If the environmental concern possibly affected citizens, her team tested blood and urine samples. In most cases, the results were given to a consulting company and not directly to the citizens themselves. Over time, this began to concern Wilma, who decided to leave the research institute and venture out on her own.

She created a business to help and assist citizens who were often neglected and had no representation. It became her personal mission "to provide citizens with the same technical expertise that the other side had," she says. Opening her own consulting business and laboratory wasn't easy. The greatest challenge, she says, was getting a loan from a bank. Several banks turned her down. But she persevered and eventually found a bank that granted her the money to start her business. Fifteen years later, her business is growing stronger than ever.

What do you do all day?

Wilma's business is quite unique in that she is a consultant with her own laboratory. On some days, Wilma is in the lab doing testing. Other days she's in the field taking samples or meeting with citizens, business owners, government legislators, or regulatory agency officials. In addition to her citizen clientele, she has many businesses as customers. They hire her for environmental consulting and technical assistance. A business owner may seek her advice on how and where to dispose of waste or may need advice on meeting government environmental codes. Her company also monitors wastewater discharge.

However, the bulk of Wilma's work is with community residents affected by an environmental dilemma or threat. One typical case concerned an oil-waste site where toxic liquids were flowing into nearby streams and contaminating water wells. She says the state agency wasn't helping the people, so she was hired. Wilma collected samples and data. She then showed the data to the state agency, pointing out where and why there was a problem.

Wilma's technical expertise encompasses only one part of her job. Equally as important, she acts as the citizens representative to persuade—sometimes even do battle with—government agencies, corporations, or landowners to create change. There have been cases where the community was so poor that Wilma, out of her deep sense of commitment, has done the work for free. She also lobbies foundations and other resources for money to cover those costs.

Where do you see this job leading you?

Wilma finds the work she's doing fulfilling. She feels fortunate to have successfully created a career that allows her to do the work she loves, and to make a good enough living to be able to grant her services for free to people who are cash-poor but need help in overcoming an environmental threat.

> **LEARN BY JOINING A COMMUNITY GRASSROOTS ORGANIZATION.**

86. Fund-Raiser

description: It takes money to get the job done. Most of the large, well-known national environmental organizations employ full-time professional fund-raisers to secure endowments.

Depending on the size of the organization, fund-raisers' jobs may be specialized. For example, some organizations may have fund-raisers whose job it is to solicit grant money from the government, large corporations, or philanthropic foundations. There are some fund-raisers who specialize in what are called "major gifts"—they seek large donations that can range from thousands to millions of dollars. Other fund-raisers may rely strictly on reaching out to the general public for smaller handouts. Their duties might include the supervision and coordination of direct mail and telephone appeals.

Whether it's a $5 donation or a million-dollar gift, these contributions are the backbone of a nonprofit organization's budget and of its success in its mission.

salary: Entry-level salaries for professional staff fund-raisers range from the low teens to $30,000 yearly, and more experienced, high-level fund-raisers can earn from $30,000 to $100,000 yearly.

prospects: Currently, there's a lot of opportunity in this field. This profession has experienced tremendous growth in the last 15 years, mostly because federal government budget cutbacks have forced nonprofit groups to rely on their own resources to find money.

qualifications: In recent years, there's been a strong effort to establish professional standards within the field of fund-raising. Employers are looking for certified, experienced, and qualified fund-raisers. Certification is offered by the National Society of Fund Raising Executives. Business and finance backgrounds are helpful, but are not always required. One instead must show proven experience and ability to successfully solicit donations for an organization.

characteristics: First and foremost, a fund-raiser must be personable. To be a successful fund-raiser, you must also have a clear understanding of the mission and issues that the particular organization tackles. You also need somewhat of an entrepreneurial spirit to achieve the financial goals required to meet the organization's operational needs.

Ruth Swanson *is a fund-raiser for a national nonprofit association that specializes in outdoor issues.*

How did you get the job?

Ruth Swanson began her career far from the world of environmental advocacy; she worked as a tour manager of an opera company. When the company was on hiatus, Ruth looked for a temporary job, and found one for the Sierra Club. After a few months as a "temp," Ruth got a permanent staff position in the corporate foundation relations department, where she began working as a grant and report writer. She eventually moved up into the fund-raising position she has now, where she specializes in acquiring large donations.

Ruth says she got the job through hard work and an attention to detail. Even though she didn't have any direct fund-raising experience (she has degrees in business and music), her managerial skills from her days at the opera serve her well in this career. Ruth is now a regional director of major gifts. "I am primarily responsible for identifying, cultivating, and soliciting major gifts from individual donors." She coordinates an area called "planned giving,"

which means getting people to designate the organization as a beneficiary in their wills or life income trusts. She also does foundation fund-raising and coordinates major phonathons and direct mail appeals.

What do you do all day?

Fund-raising is a slow process. Getting grants from foundations can take months if you take into account the stages from research, to writing proposals, to when you actually hear a reply. But Ruth spends most of her time asking longtime members to donate substantial amounts. She also runs special fund-raising projects.

One recent day, Ruth glanced at her to-do list and found that she had to coordinate a campaign to collect $75,000 over six weeks. She's now interviewing volunteers and setting goals.

Ruth spends a lot of time on the phone and on e-mail. She phones the organization's longtime members. "If their immediate reaction is 'Oh no, a

fund-raiser,' I will try to change their mind about me and break down the bad image they may have about fund-raisers. If they say, 'I hate being called by fund-raisers,' then I'll write them a card apologizing for the inconvenient time I called, but I'll explain what we're doing and give my phone number for them to call me."

Because she is in charge of a large region that covers several states, Ruth travels about one week a month. She meets with local association chapters and spends time with individual members in their homes, explaining issues and strategies and why the organization needs money to tackle these topics.

> **LEARN BY VOLUNTEERING. ENVIRONMENTAL GROUPS WELCOME PEOPLE WILLING TO HELP THEM RAISE MONEY. START WITH A SMALL GROUP OR OFFER TO WORK ON A PART-TIME BASIS.**

Ruth says the end of the year is usually the busiest because that's when people tend to donate. The spring, however, is when you begin cultivating contributors by educating them about your group's issues and causes, so that by the end of the year, your work comes to fruition.

Does she ever feel like a salesperson just asking for handouts? "No, I've never felt that way because I've never pushed anyone into making gifts that they didn't want to make. I feel that fund-raisers have an incredibly high moral obligation to honor the requests and the interest of the people out there who are making gifts. I've never felt uncomfortable about what I do. In fact, I feel quite proud of it."

Where do you see this job leading you?

Ruth is very happy working where she does. "It's home," she says. "There are lots of different causes you can relate these skills to—for example, the Girl Scouts or public television. In terms of career options, it's a really good one."

87. Epidemiologist

description: Environmental epidemiologists are scientific medical detectives. They investigate how exposure to particular environmental contaminants affects your body and causes disease or disability. Most environmental epidemiologists work for federal, state, or local public health departments. Jobs can be found with the U.S. Department of Health and Human Services, the Centers for Disease Control and Prevention, the Agency for Toxic Substances and Disease Registry, the National Institute of Environmental Health Sciences, and the Environmental Protection Agency. Many also work in medical centers and universities, while some others are teachers or researchers at universities or schools of public health.

On a daily basis, epidemiologists field calls from community residents and medical personnel concerning clusters of diseases or ailments that may warrant inquiry. At the start of each new investigation, the environmental epidemiologist usually spends time in the community taking samples and conducting interviews. Like detectives, they track any clues or leads that will help them solve the mystery. They conduct laboratory tests, analyzing samples and other data. Once they uncover the cause, environmental epidemiologists offer diagnosis and/or possible treatment, or recommend environmental changes to remedy the crisis, and to prevent further illness or disability.

salary: In the private sector, those with master's degrees earn between $40,000 and $50,000. Those with a medical degree and/or public health degree earn $60,000 to $90,000. In the federal government, entry-level salaries for those with master's degrees are between $27,000 and $40,000 a year.

prospects: There are terrific opportunities in environmental epidemiology. While jobs in infectious diseases epidemiology are getting harder to find, the job market for environmental epidemiologists is booming.

qualifications: At minimum, environmental epidemiologists need a master's in epidemiology or public health. A doctorate is recommended. Your background should include environmental science courses.

characteristics: You have to love solving mysteries. Inquisitiveness, thoroughness, and persistence are the three most important characteristics needed to be a successful and effective environmental epidemiologist.

Dr. Ruth Etzel *is a supervisor of environmental epidemiologists at the Centers for Disease Control and Prevention in Atlanta, Georgia.*

How did you get the job?

Ruth Etzel was a practicing pediatrician when she became increasingly aware of how environmental factors affected the babies she was seeing in her clinic. That interest led her to enroll in the school of public health that was just across the street from the hospital where she practiced. After getting a Ph.D. in public health, Ruth was accepted into the two-year training program at the Centers for Disease Control and Prevention (CDC) in Atlanta. Her job as an environmental epidemiologist was to track down diseases and find their causes. "I decided it was the perfect job for me because I love mysteries," Ruth says. When she finished the training program, she stayed on with the CDC and is now the Chief of Air Pollution and Respiratory Health Branch, at the National Center for Environmental Health, Centers for Disease Control and Prevention.

What do you do all day?

It depends on the issue of the day. At any time, Ruth's office may get a call about an epidemic in the United States, and sometimes outside the country. She will coordinate a team of environmental epidemiologists to travel to the affected city or town. Within 24 hours, the CDC can have a team on site, collecting samples of data and conducting interviews. For example, Ruth had just returned from one city where doctors had noticed an increasing number of infants becoming seriously ill and coughing up blood. She and her team met with doctors and conducted extensive interviews with parents. They also collected samples from the homes. Back at the CDC's offices, they analyzed the information and worked to solve the mystery as quickly as possible.

In some cases they are able to uncover the cause within days. Sometimes it can take months or years. As quickly as they discover the cure, the team begins

> ONE ENVIRONMENTAL EPIDEMIOLOGIST SUGGESTS THAT EPIDEMIOLOGY GRADUATE STUDENTS CHOOSE A DISSERTATION TOPIC THAT INVOLVES THE ENVIRONMENT. THAT WILL DEMONSTRATE TO PROSPECTIVE EMPLOYERS THAT YOU ARE SINCERELY INTERESTED IN THIS FIELD.

to formulate possible solutions and to send out information about how to prevent the problem.

Ruth says that during a typical day at the office, she spends 10 percent of her time on the phone talking with doctors, 20 percent on the computer analyzing data, and 20 percent talking with public health educators and others about possible interventions. The rest of the time, she writes and presents reports to advise communities and doctors on how to handle environmental health problems.

Where do you see this job leading?

Ruth doesn't want to go back to pediatric practice. "Once you've gotten to see the community as your patient rather than the individual, you feel so much more satisfied solving the community's problems, because you've got a lot more impact."

88. Industrial Hygienist

description: Industrial hygienists are scientists and engineers committed to protecting the health and safety of people in the workplace and the community. Members of this profession help identify and control environmental factors that may result in injury. Industrial hygiene is considered a science, but it also involves judgment, creativity, and human interaction. Typically, industrial hygienists work with issues such as lead exposure, asbestos abatement, indoor air quality, noise, ergonomics, sick building syndrome, radiation, and exposure to chemical and physical contaminants. They also ensure that safety laws and regulations are being followed. This job can found in public utilities, government agencies, research laboratories, hospitals, hazardous-waste companies, chemical companies, manufacturing plants, insurance companies, and others.

salary: Salaries depend, of course, on the type of industry you choose to work in. However, according to the American Industrial Hygiene Association, entry-level industrial hygienists are generally paid between $28,000 and $35,000 a year. Mid-level industrial hygienists make about $40,000 to $70,000, and the top-level salaries are $60,000 and higher.

prospects: A 1993 *U.S. News and World Report* survey ranked industrial hygiene as one of the top professions of the future. *Working Woman* magazine recently identified industrial hygienists or environmental managers as one of the top five "cutting edge" careers. Unlike many other professions, the position of industrial hygienist is not limited to one particular type of industry. For that reason there are many employment opportunities.

qualifications: Though some colleges and universities offer undergraduate programs in industrial hygiene, students usually prepare for this career by pursuing an undergraduate degree in a science such as engineering, chemistry, or biology, and then earning a postgraduate degree in industrial hygiene. Some colleges offer a one- to three-year associate degree and certificate program that qualifies students as industrial hygiene technicians. Technicians assist in analyzing data and ensuring programs and regulations are enforced. After working in the field for five years, you are then eligible for industrial hygienist certification by taking a comprehensive two-day exam.

characteristics: This career requires someone who is an analytical thinker, who has a high level of technical curiosity, and who has strong people skills.

Fred Freiberger *is an industrial hygienist with a manufacturing firm.*

How did you get the job?

Fred Freiberger didn't start out wanting specifically to be an industrial hygienist; he was originally a manufacturing engineer. He first became interested in this career when he was manufacturing aerospace hardware. As part of his job, he visited the space program facilities, and there became intrigued with the focus on safety. He later became a loss prevention engineer for an insurance company. In that role, he visited policyholders. "I visited virtually any business or enterprise you can imagine, from industrial chemical operations to public parks to cosmetics and pharmaceutical manufacturers," Fred says. He advised them on ways to prevent workplace-related illness or injury. He then moved to IBM, where he worked for 20 years. He eventually created his own consulting business and then heard about the opportunity with the manufacturer where he now works.

> HAVE A CAREER PLAN. ALTHOUGH A CAREER PATH CAN ENDURE TWISTS AND TURNS, ONE INDUSTRIAL HYGIENIST COMMENTS, "IF YOU DON'T KNOW WHERE YOU'RE GOING, THEN HOW ARE YOU GOING TO GET THERE?"

What do you do all day?

As a safety engineer, Fred keeps people and processes safe in a large multiplant corporation. His job is to oversee the safety of everything from small office equipment to three-story-high printing presses. He also checks supplies, materials, and chemicals for flammability and toxicity. Fred says he spends a third of his time at his desk pushing paper, preparing hazardous material data sheets, or responding to internal questions. Another third is spent walking around the plant checking day-to-day operations. The rest of his time is spent on employee training.

That training, he says, is ongoing. "Whenever we have a new product or new process, one of the requirements is that it come through the safety department for review. We review it on the basis of the materials that are used, the equipment, and its effects on the employees and environment," he says. Recently, Fred was conducting fire-safety training seminars. Training days are long ones, as he must stay and train employees on each of the three shifts. He often has to travel to the company's other plants, which entails a two- to three-hour drive.

Some of the changes that Fred has made at the company include increasing the number of mechanical lifting aids, such as power lifts, hoists, and forklifts, to help employees move heavy materials. He also suggested discontinuing use of flammable solvents to mop the floors. In other areas, ventilation was improved based on his recommendations.

Where do you see this job leading you?

Fred has been with this company only a few months, and he feels he's just getting started making improvements with this manufacturer. However, he says, as the world changes, the industrial hygiene profession is constantly faced with new challenges. The American Industrial Hygiene Association says there's a wealth of opportunity in this profession as managers realize that industrial hygienists help increase worker productivity and reduce insurance and medical costs, thus lowering overall costs while keeping workers safe and healthy.

89. Meteorologist

description: Predict the weather—that's what a meteorologist does. Meteorologists use atmospheric science to forecast weather patterns and anomalies as accurately as possible. A meteorologist's mission is the protection of life and property by warning the public of impending weather hazards. Data is collected from ground weather observation stations as well as from satellite stations. Meteorologists print out maps and analyze such variables as barometric pressure, wind, humidity, and precipitation. Using computer weather models, they are able to forecast changes and decide if special weather alerts should be issued to the public. There are several forecasting areas in which to specialize, such as severe storm or tropical weather forecasting.

salary: Salaries usually start at around $20,000 a year, but some of the top, more experienced forecasters can make somewhere in the mid-$50,000s.

prospects: Right now, job prospects with the National Weather Service are not good. However, state and county governments hire meteorologists, and so do many private corporations. There are also privately owned commercial meteorological services that provide forecasts to television and radio stations as well as to shipping and aviation concerns and private industry. Your chances will be better if you try for a job in the private sector.

qualifications: A degree in meteorology is required.

characteristics: A meteorologist must be self-motivated, a critical thinker, and decisive. He or she must also be tolerant of working long hours, overnight, and weekend shifts.

Matthew Tauber *is a meteorologist for the National Weather Service.*

- -

How did you get the job?

It wasn't easy for Matthew Tauber to get his current job. He waited three years from the time he applied with the National Weather Service to when he actually got hired. As luck would have it, just as Matthew was graduating from college, the National Weather Service instituted a hiring freeze. With a degree in meteorology, he found himself working odd jobs until the Weather Service finally called him. He says he could have applied with commercial meteorological services, but because his dream was to work for the National Weather Service, he decided to stick it out on the waiting list. His patience paid off.

What do you do all day?

You certainly need to be flexible to hold this job, which can wreak havoc on your body clock. Matthew's eight-hour shift changes every week; one week it's the day shift, the next it's the afternoon/night shift or the overnight shift. The staff also rotates working weekends. Keep in mind that the Weather Service is open 24 hours a day, seven days a week, including holidays. When there is severe weather brewing, the hours can be long and the work grueling. You may have to work overtime or get stuck working a double shift if your relief can't make it in because of the very same bad conditions. Even in the calmest weather, the work is never routine.

On a rotating basis, Matthew is assigned to be either the aviation forecaster or the regular public forecaster. Most of the time, he's at the computer checking readouts from observation stations around the country. He prints out maps and analyzes them for current conditions such as barometric pressure, wind, humidity, and precipitation, as he calculates and develops short- and long-range forecasts. The room is

> **YOU MUST BE WILLING TO RELOCATE AND WORK WHERE THERE'S A JOB OPENING.**

filled with computers, some of which are satellite computers on which the weather imagery transmissions are received on a continual basis, while others are those on which the forecasts are disseminated. When Matthew's not at the computer, he's scanning radar screens.

But as a government employee, he's also a public servant. Part of the job means answering phones and telling callers the forecasts. The upside of the job is that he gets satisfaction from the fact that his weather predictions have an impact on people's daily lives. The downside of the job, Matthew says, is the abuse from the public. "You put up with a lot [of criticism and jokes] . . . you have to just put up with it," he says.

Where do you see this job leading you?

This is Matthew's dream job. As a child, he was fascinated by the weather, especially thunderstorms and snowstorms. When he found out that people actually make a living at it, he wanted to try it himself. Now that he's achieved his goal, he wants to stay right where he is. It is possible for him to move up the ladder to lead forecaster, which is a supervisory position. However, a bit of uncertainty looms over the National Weather Service, as federal budget cuts threaten a downsizing of the staff. There's also been talk of contracting out some of the services to privately owned weather-predicting companies.

90. Ecotourism Salesperson

description: Ecotourism is a new field, and the definition is about as formative as the industry. But anyone in this innovative field is bound to say that the emphasis in ecotourism is on fostering respect for different cultures and environments. People taking a trip set up through an ecotourism agency should be prepared to immerse themselves in a new culture and experience life through a different mode. Five-star hotels and steak dinners most likely won't be part of the game plan. An ecotour is more likely to include white-water rafting and a night on the riverbanks.

Though ecotourism salespeople spend much of their time on the phone selling tours to clients, they also have the occasional opportunity to find and test new tours.

salary: A typical starting position in an ecotourism company would be that of a salesperson. Depending on the company, new salespeople either will be paid about $8 an hour or will receive a commission from their sales. A salesperson working on commission will bring in anywhere from $1,200 to $3,000 a month. Top jobs in this field, such as a vice presidency in sales and marketing, can pay anywhere from $40,000 to $75,000 a year, depending on the size of the company.

prospects: These types of jobs aren't abundant because there aren't a large number of ecotourism companies out there. But the number is growing, and consequently, more opportunities are becoming available.

qualifications: A college education is not required for someone wanting to get into a sales position in ecotourism; however, for anyone who desires to progress into a more supervisory role and bring in a larger salary, a bachelor's degree is recommended. A degree in marketing is ideal for this type of work.

characteristics: People in ecotourism sales are personable, have a high level of energy, and, for the most part, enjoy being outdoors. Employers in this field are often looking for someone who is detail-oriented, self-motivated, and able to absorb new information quickly.

Marc Ahumada *is an ecotourism salesperson.*

How did you get the job?

When Marc Ahumada was a young boy, his father owned a travel agency that organized tours into Mexico and Hawaii. The travel industry would have been a natural business for Marc to go into, but he wanted to try something different. So when he went off to college, he got his degree in marketing and fashion merchandising. Marc was mostly interested in active wear, more specifically surf wear, and for five years after college he worked for various surf wear companies in their marketing and design departments.

"Fashion merchandising is really competitive and it's really cyclical. One of the companies I was working for laid off a bunch of us because their sales went down, and we were the first to go," Marc says. He then decided to go into business with his father, who had since closed his travel agency and was working as a consultant. The two men added a third partner, a long-time friend of both Marc and his father, and they opened TourTech International.

Ten years into the company's life, Marc says, he couldn't be happier with his choice to get into the travel industry. "A lot of my friends think that I have the greatest job in the world, and I tend to agree with them," he says.

What do you do all day?

For the most part, Marc's job consists of travel. He spends at least one week of every month out of the office—usually out of the country. During the busy seasons, which are spring and fall, Marc may spend only one week of every month in the office.

While he enjoys the travel, he says travel for work is different from vacation time. "In that short time you have out of the office, you have to see as much as you possibly can. You're not taking a leisurely pace about it. You're driving, you're stopping, you're looking at a hotel, you're seeing what the amenities are there. You're looking at a lodge, you're seeing where it's located, what kind of animals are there, and what kind of ecosystem it's in. You're going to a beach area and you're seeing what types of waves there are and what kind of activities can be done from there." There is also a lot of preparation that takes place before a trip, and Marc says that sometimes that is what makes the travel so enjoyable.

"I love traveling to these places and discovering all the new things that there are to discover, and reading up a storm before I go, then going and actually seeing the things and tasting the food."

During the time Marc is in the office, he spends a lot of time with the negotiations of new travel packages or training the sales staff. He also devotes a lot of energy to the marketing and sales aspects of the company. Putting together brochures, videotapes, and information packages about the different tours is a large part of this work.

> BECAUSE THIS FIELD IS SO NEW, YOU WILL BE HARD-PRESSED TO FIND ANY TYPE OF ORGANIZATION THAT PRODUCES LITERATURE LISTING THE DIFFERENT COMPANIES OR AVAILABLE JOBS. THE BEST THING TO DO IS LOOK THROUGH THE YELLOW PAGES UNDER TRAVEL AGENCIES.

Marc also does a lot of speaking, both to the public and to members of the travel industry.

Where do you see this job leading you?

Marc sees himself as eventually becoming the president of TourTech International.

91. Ecotour Guide

description: An ecotour guide escorts paying tourists to remote, sometimes environmentally sensitive areas so they can see and understand the wildlife and topography of the area. The guide is responsible for knowing where to go, how to get there, and where to stay. Ecotour guides specialize in regional travel, wildlife photography, or bird-watching. It's a job for people who love to travel and don't mind being away from home for long stretches of time.

salary: At well-established companies, tour guides will begin at a rate of $100 a day, but they are seldom hired as salaried staff until they've proven themselves valuable to the company. Experienced full-time ecotour guides will earn $30,000 to $35,000 a year.

prospects: Jobs are limited and competition is tough, but there is a sense in the industry that things are expanding. Lots of small outfits have been opening up in the United States and Latin America, while Europe has seen a surge in demand for bird-watching guides. For qualified people, it is possible to get a foot in the door.

qualifications: Many ecotour companies will look first for credentials: an advanced university degree, authorship of books or papers on the relevant wildlife or region, or any other formal demonstration of your expertise. It is essential that you be able to display detailed knowledge of the subject matter. For instance, anyone seeking to conduct bird tours should already have a working knowledge of ornithology before attempting to land a job. "Birding," as enthusiasts call it, is not something you'll be able to pick up after a few weeks of intense study. You must be better informed than the people booking the tour. In short, you have to be an expert.

characteristics: Even with all the knowledge in the world, you will be unable to succeed in this field without "people" skills. Remember, you will be in charge of groups of people with nothing in common but their love of birds, wildlife, or travel. It's your job as an ecotour guide to make sure they enjoy themselves and come away with a memorable enough experience to justify the money they've paid. In this job, you will need to be able to think on your feet. Imagine arriving at a hotel in a remote area and finding that there are no reservations for you or your customers, or having to administer first aid to a customer with a broken hip or a poisonous snakebite; those are the kinds of challenges you can expect.

Barry Zimmer *is an ecotour guide operating out of El Paso, Texas.*

How did you get the job?

Barry Zimmer has always been interested in birds. He has been "birding" since the age of eight, but never thought he'd make a living at it until opportunity knocked. He was finishing up a psychology degree with a minor in biology, when a friend whose parents owned a travel company asked him to escort two women who wanted a private tour of Texas. He accompanied them from Houston down the Texas coast to see the birds migrate north from Mexico. That got his foot in the door. Then, just as he was about to graduate, another tour guide got sick just before a trip to Arizona. Barry was asked to step in with one day's notice, and he pulled it off. His career was under way.

> IF YOU HAVE EXPERTISE IN A PARTICULAR AREA, TRY TO DEVELOP A REPUTATION AND PROMOTE YOURSELF TO TRAVEL COMPANIES. YOU WILL NOT SEE THESE JOBS ADVERTISED.

What do you do all day?

When not conducting a tour, Barry is preparing for one. There are itineraries to be written as well as articles for a monthly newsletter. He also must take care of tour financial reports, since on the tours he is responsible for large amounts of money. On the excursions, everything is paid for in advance: hotels, food, and transportation. Once he's returned from a trip, Barry lists all the species of birds, mammals, and reptiles sighted on the trip and sends those lists to the sightseers. When on the road, he is usually riding in one of the two large passenger vans that travel through the planned route. Trips last from five days to three weeks at a time. Barry handles trips throughout the United States, including Alaska. Foreign travel includes Canada, Mexico, Belize, Costa Rica, and Antarctica. He is on the road as much as 140 days a year.

As a tour leader, Barry says, he does everything from driving the van, to preparing picnic lunches. Although he has to know his subject, a lot of what he does has nothing to do with birds. He says the frustrating part of the job is when people hand him a list of birds that they demand to see. They'll pressure him, saying, "There's only four birds that I want to see and I've spent $4,000 to come on this trip. I want to see them!" But, Barry says, most people aren't like that.

Barry says he never tires of returning to the same locations or seeing the same birds. (He's taken people to Arizona 20 times, for example.) "What's really neat is seeing other people see certain birds for the first time," he says. "There is a huge responsibility to take care of the group, the pressure of finding the birds, and that sort of thing, but you are getting paid to be out in nature and do what you love to do the best." He recalls one trip especially fondly: "I'm in Antarctica and I'm standing around with 60,000 penguins and I'm getting paid to do this."

Where do you see this job leading you?

Barry says without hesitation, "This is definitely it for me." He says he's in a good situation with the company he works for, including benefits. "The only downside is the time I spend away from my family," he says. "This is the perfect job for a bachelor."

193

92. Eco-Fashion Designer/Entrepreneur

description: The "natural" look is in. All-natural, hemp-colored bed linens and woven cotton blankets are extremely popular. A growing concern for the environment prompts consumers to look for "green," or environmentally friendly, products. One emerging new market is the fabric industry. This involves making clothing or linens from unbleached, environmentally safe yarns. The simplicity, feel, and designs of these fabrics are quite appealing to the increasing number of environmentally aware consumers. There are designers and retailers promoting and benefiting from this environmental interest. Eco-friendly products are no longer a specialty, but are being marketed to the mainstream.

salary: You may start with nothing but an idea and a loan from the bank, but the profits are unlimited.

prospects: The prospects are wide-open if you want to start your own eco-fashion business. However, you may first want to gain experience by getting a job with an established designer or retailer. In that case, the prospects for employment may be somewhat limited, because there aren't that many companies creating these garments. As the trend for these clothes heightens, watch for garment companies jumping on the "natural" bandwagon.

qualifications: A knowledge of fabric and design is recommended, as well as a background in art. Business courses are also helpful.

characteristics: One must be creative, aggressive, original, persuasive, and motivated. A designer working for a company must be able to interpret a chief designer's ideas and translate them onto paper and into patterns, production systems, and garments.

Marylou Sanders *is an eco-fashion designer and co-owner of her own eco-fashion design company.*

How did you get the job?

Marylou Sanders has carved her own career path. Not satisfied with a degree in fine arts, she went back to school to study fashion design. Eventually, she and her husband, Daniel, launched a successful T-shirt company. One day, while thumbing through a Greenpeace catalog, Marylou noticed that the T-shirts they sold were not 100 percent cotton. She thought that it was ironic that the T-shirts were half polyester, an oil-based product, while Greenpeace was battling oil companies. She brought this to the attention of the merchandiser for Greenpeace, who said, "Give me an alternative."

Marylou met the challenge, tracking down organic cotton farmers and purchasing as many bales as she could afford. That's

> **CHECK DESIGN SCHOOL JOB BULLETIN BOARDS.**

when she ran into the first major obstacle—organic cotton is not as processed as commercial cotton, and is therefore much more difficult to spin into yarn. Marylou hired an expert cotton converter, who engineered a new spinning method.

Their first major sale of fabric was to Esprit, a top clothing manufacturer. Marylou says she and Daniel started with $10,000. In only five years, their company has become a $6 million business. Marylou says they not only created a family business, but also are responsible for spurring on the entire organic cotton industry. "No one was doing anything in this field," she says. Now there are more farmers growing organic cotton than ever before. They not only launched a new technology, "we spurred the industry and influenced fashion."

What do you do all day?

Marylou is always working on two or three things at once. While she was on the phone with me, she was penciling designs. As we talked, she was interrupted by an assistant bringing in new color cotton swatches for her approval. Marylou not only oversees the manufacturing of yarns and fabrics that are sold to garment manufacturers, but also designs her own line of casual sportswear for men, women, and children, called Ecosport. In the next room, there are three people sewing, plus a pattern maker and a cutter. With two young children, Marylou is used to juggling several things at once. When she's not overseeing designs, she is checking the warehouse where shipments are packaged and mailed.

If that wasn't enough, last weekend Ecosport opened an outlet. "After I get off the

phone with you, I'm going to a place to buy zippers. I have to stop by the outlet and deliver a cash register. Later I'll come back and work on designs again to check fit and test shrinkage. I have a meeting about the fall designs. Then at 4:30, when the kids come from school, I turn into 'Mommy.'"

Where do you see this job leading you?

Marylou looks ahead to expanding the company. Organic clothing, she says, isn't limited to colorless fabrics. They're now experimenting with exciting natural dyes to create colorful fabrics and designs. She says there are so many more eco-sensitive product lines in which to venture: "I see only openings where many people might only see doors."

93. Eco-Entrepreneur

description: Take a great idea, a commitment to the environment, and the tenacity of someone who won't take "no" for an answer, and you've got an eco-entrepreneur. Consumer demand for "green" products—that is, environmentally safe products or products that promote environmental concerns—is growing. Every aspect of the growing environmental consciousness represents a potential business, and those with an entrepreneurial spirit can jump on the bandwagon and make money from it. There are different types of eco-entrepreneurs: those who create, package, and sell products that are environmentally safe, and those who create products, systems, machinery, or technology with an ecological purpose. Examples include a greeting card business that prints on recycled paper, a new recycling machine, or an environmental game or toy. Like entrepreneurs in general, an eco-entrepreneur takes an idea and researches and fine-tunes it. The next steps are creating a business plan, seeking financing, and either selling the idea for top dollar or launching a business. The possibilities are as unlimited as one's own imagination.

salary: Salary potentials are unlimited, and equally as undefined. Basically, you get out what you put in.

prospects: There continues to be an emerging market for products that are safe for the environment, our general health, or the health of workers. Essentially anything that involves an environmental component has the possibility of selling. If it's a good idea, it will sell. The tough part may be convincing those who are in the position to finance or mass-market your idea.

qualifications: This career path requires a relentless entrepreneurial spirit with knowledge, research, and commitment to support one's idea or product. Part of that research may involve taking college science or business courses. In addition, an ability to write is essential for proposals.

characteristics: Creativity is the number one characteristic you'll need. First, you have to come up with a great idea. Persuasiveness is number two, as you have to exude a sales mentality to get your idea sold. You'll also need to be able to articulate your idea to financiers as well as to the buying public. You need the sense of professionalism to follow through on the idea. As one eco-entrepreneur says, "A lot of people have great ideas, but they don't necessarily have the ability to piece [them] together and come up with a product that is professional."

Michael Stusser *is an eco-entrepreneur and the creator of Earth Alert, the Active Environmental Board Game.*

How did you get the job?

This is not a job you get—it's one that you create. In 1990, Michael Stusser and a friend came up with the idea for Earth Alert. The idea didn't just come in a flash. He and his buddy brainstormed, tossing around several ecology-oriented ideas. "We asked ourselves, what about eco-coffee or eco-pinball?" Michael laughs. Finally, they came up with Earth Alert. The board game requires players to answer trivia questions and act out charades. Michael's hope is that the game is a fun way for players to learn more about the planet and help to save it. As part of the game, players might draw a card that requires them to perform an energy-saving action such as turning off a light in the house or writing a letter to the president. Major toy companies said

the game would never sell and advised Michael against it. Not taking "no" for an answer, he and his partner did all the research and product testing to get the financing that would eventually allow them to market and launch the board game themselves. To date, they've sold 30,000 games.

Michael's strong affection for environmental concerns dates back to his childhood in Seattle and the weekend hikes he took with his parents. In college he majored in journalism and political science, but quickly decided he didn't want to be a reporter. He answered a newspaper ad and got his first job canvassing door to door for a public interest research group. He did that for a summer and, although it paid very little, it led to a promotion. Before he knew it, he was working as an organizer for the group. That led to other jobs with environmental organizations as a media relations coordinator and lobbyist.

What do you do all day?

Michael works out of his home. Every day he's on the phone lining up distribution and new clients and calling media contacts. Recently, he was preparing for September, the busiest time of year, when he gears up for Christmas. He was arrang-

ing promotions, including an appearance on the *Today* show. Michael is also an educational game creator for Entros, a company that markets interactive games for museums, schools, and science centers.

Where do you see this job leading you?

"I would hope more of the same. The opportunities are endless. The environmental crisis is getting worse. There are so many issues we need to tackle. I see what I do as educational outreach in innovative ways."

> **START BY FORMULATING AN OVERALL PHILOSOPHY OF YOUR IDEA AND MISSION. IDENTIFY THE MARKET FOR THE PRODUCT OR SERVICE. FOR FINANCING, SEEK OUT INVESTORS OR FUNDS SUCH AS THE GLOBAL ENVIRONMENT FUND, WHICH INVESTS IN ENVIRONMENT-ORIENTED BUSINESSES.**

94. Project Manager

(ENVIRONMENTAL CONSULTING FIRM)

description: This is one of the fastest-growing industries in the country. There are environmental consulting firms advising private companies and the government on every aspect of the environment imaginable. Consulting may be one of the single most popular fields for environmentally oriented college graduates. One of the reasons is money; another is corporate advancement. In some firms, entry-level is a field technician job, which involves traveling to project sites to take samples and collect data. A field technician also might examine company records, regulations, and licensing requirements. The field technician works under the direction of a project manager. It's the project manager's job to take the client's directive, research the task, and get the job done. The project may require working with and coordinating teams of specialists and engineers from many departments.

salary: Entry-level salaries begin in the mid- to upper $20,000 range, but you can quickly make more money by advancing through a company or, better yet, jumping from one company to another. Project managers start at $30,000 and up per year.

prospects: Prospects are good, as the environmental consulting field continues to expand; however, it is very competitive.

qualifications: Degrees in any of the sciences, such as geology, biology, wastewater engineering, or environmental science, are recommended. A keen understanding of regulation is an advantage in getting ahead. Good reading skills and an understanding of scientific terminology are also required. In this job, you'll probably have to do research outside of your area of expertise. One project manager advises, "You've got to be ready to pick up the research, even if the subject is outside of your specialty, and interpret it quickly." Good writing skills for composing reports and memos are also essential.

characteristics: This job requires communication, organization, and management skills in order to implement, delegate to, and supervise work teams. It also takes poise and diplomacy to deal with demanding clients. You must be articulate enough to explain your findings and recommendations to clients and regulators.

Katherine McGillis *is a senior project planner for an environmental consulting firm.*

- -

How did you get the job?

Katherine McGillis started in the business by working for a government regional planning agency. She had just graduated with a bachelor's of science in geology, and knew only that she wanted to do something concerning the environment. She worked in the area of groundwater protection and helped implement protection bylaws. Katherine says it was great experience because it taught her about the regulatory side of the industry and also offered exposure to the different layers of government. She wanted to gain the government work experience before moving to work in the private sector; that's just what she did. She found her next job in the newspaper want ads. Working for a private consulting firm, she worked on pipeline projects and wrote environmental reports that were submitted to federal agencies. This involved taking data from maps, site visits, and research and translating them into the correct format. She did that for five years before jumping to the firm where she now works.

What do you do all day?

"I'm the client advocate, the liaison between the client and federal, state, and local regulators," says Katherine. For example, a client may want to build something in a location in which the government says getting permits will take nine months. However, the client wants to start building in three months. It's her job to speed up the process. A project manager must not only make suggestions to the client, but must also be persuasive, supported by data and research, to cut through government red tape. Yet through the process, it's the protection of the environment that remains the priority.

Recently, Katherine was working on a project for a company that wants to build recycling plants. She phoned the local Department of Environmental Protection to schedule a meeting, and then had to prepare material for the meeting.

> **DON'T BE AFRAID TO SEND A RESUME TO A COMPANY THAT IS ADVERTISING FOR A POSITION OTHER THAN WHAT YOU'RE LOOKING FOR. THE RESUME JUST MIGHT GET SHUFFLED TO ANOTHER DEPARTMENT. HOWEVER, FOLLOW UP WITH A LETTER OR PHONE CALL.**

This involved checking with all the engineers, scientists, and technicians on the project. Together they would go over the survey work, data concerning drainage issues, and regulations with regard to groundwater discharge, as well as other concerns.

Katherine usually works a regular 9-to-5 day, but she also has to travel several times a year, meeting clients and checking sites firsthand. She says the project manager is expected to be an expert about a lot of things. Although she has a degree in geology, most of the time Katherine has to tackle subjects outside of her expertise and handle them with adequate knowledge to make the correct judgment calls on behalf of her client.

Aside from the work itself, it can be a challenge to deal with difficult clients who don't follow your recommendations. That's when, Katherine says, you put everything in writing and safeguard yourself with a memo trail. If they ignore your advice and make the wrong decisions, you can pull out a memo and say, "Told you so."

Where do you see this job leading you?

As a mid-level manager, Katherine has moved up quickly, proving that a master's degree is not absolutely required for the job (although it might help win a higher starting salary). She wants to continue to advance within the corporate structure. This is an area that is beginning to open up for women. From Katherine's perspective, she sees more women working for the government as regulators.

95. Pollution Compliance Manager/Environmental Manager

description: A pollution compliance manager helps industrial polluters stay within the guidelines of the permits they receive from local, state, and federal governments. The actual job title may vary, but the responsibility is to make sure plant operations don't result in source pollution that will draw the ire of regulators and possibly result in fines or legal sanctions. In this job, you will be a type of industrial speedometer, letting your bosses know if they're beginning to exceed the local limits of source pollution. You're also responsible for helping them maintain efficient operations without violating the law.

salary: Salaries range from $25,000 to $30,000 a year to start, but increase with levels of experience and advanced degrees. This is not an entry-level position.

prospects: People able to work both sides of the permit process are in strong demand. The huge Environmental Protection Agency permit program has caused a shortage of people conversant in the sometimes labyrinthine procedures that are required. If you know anything about the actual plant, you have a big jump on the competition.

qualifications: A bachelor's degree in environmental science is the absolute minimum. A master's degree will get you a job quicker and at a higher salary. More and more employers in this area are looking for people with advanced degrees. Industrial knowledge can be acquired on the job, but it helps to have an understanding of your potential employer's business needs and problems.

characteristics: You must be able to supervise others. As a pollution compliance manager, you'll be asked to direct the gathering of samples and the setting up of monitoring systems. You must have a firm grasp of the permit process, stay abreast of changing regulations, and be familiar with every aspect of your plant that could affect monitored pollution levels.

John Foley *is the environmental manager for a privately owned resource recovery plant.*

How did you get the job?

John Foley started as a "stack tester," physically climbing industrial smokestacks and pushing probes into the emerging smoke. He landed his first job out of college by enrolling in a work-study program during his final semesters. By the time he earned his bachelor's in environmental science, he already had a foot in the door. From "probe pushing," John moved on to installing continuous emission systems, setting up mobile laboratories for sample testing, and writing complex quality assurance plans as part of his employer's operating permits. It took five years to land the management job through industry contacts.

What do you do all day?

When he isn't supervising paperwork and assembling data, John dons his hard hat and monitors plant operations. His job is to make sure that the plant complies with environmental regulations regarding air emissions, solid waste, water, and wastewater. While his job now is primarily filled with managerial duties, he does routinely walk through the plant checking the monitoring systems gauges. He also must make sure that employees are doing their jobs by keeping records up to date.

> **WORKING FIRST FOR A REGULATORY AGENCY CAN GIVE YOU AN EDGE IF YOU'VE DECIDED ON HEADING INTO A CAREER AS AN ENVIRONMENTAL COMPLIANCE MANAGER.**

The environmental compliance manager or director's job is filled with deadline pressure. There are separate deadlines for the different categories such as air quality, water, and wastewater reports, and deadlines differ at the local, state, and federal levels. On top of that, John must also keep up with changing regulations and plant procedures. As he tracks changing regulations, he tells facility operators what steps are needed to comply with environmental codes. John says that although the plant where he works is relatively small, power plants are one of the most heavily regulated industries in the country.

John's plant burns trash to generate electricity. "It's not particularly glamorous," he told us, "but if facilities like this didn't exist, then people's trash would either sit on their curbs or take up diminishing landfill space. I believe it's the best disposal option."

Where do you see this job leading you?

John's not interested in starting his own consulting business because of all that goes with it, so this is the perfect situation for him right now.

96. Earthquake Safety Retailer

description: There are many areas in which someone working in the earthquake safety industry could find employment; perhaps owning your own safety equipment company would give you the widest range of experiences in this business, from inventing and manufacturing to selling products. As the owner of an earthquake safety retail business, you will spend a large amount of time marketing your product, which includes educating prospective clients as to why they need to purchase such equipment to safeguard their living or working environment against natural disaster. However, equal effort will be devoted to administrative matters. While this type of business can be exciting just after an earthquake has occurred, it can be very slow when there is not much activity underground.

salary: Because the business activity in this industry depends heavily on Mother Nature, who is never predictable, the salary for a company owner differs from year to year. In a year with a large amount of earthquake activity, you could make as much as $100,000. In other years you may make about $50,000.

prospects: The number of jobs available in this industry will depend on the business activity at any given time, and will be high following an earthquake.

qualifications: For someone wanting to start a business in this field, a knowledge of seismology wouldn't be valued as highly as good business sense. At least a bachelor's in some form of economics or business administration is recommended.

characteristics: Like owners of other businesses, the owner of an earthquake safety retail operation must have an entrepreneurial spirit. Someone in this industry must also have dedication and be able to stay in business during the slow years.

Tom Rundberg *is the owner of a West Coast earthquake preparedness shop.*

How did you get the job?

During California's 1987 Whittier Narrows earthquake, Tom Rundberg's daughter narrowly escaped injury when the entertainment center in the Rundbergs' living room swayed but didn't fall. Following this terrifying experience, Tom searched for something that would keep the entertainment center from falling during future quakes. "I went down to the hardware store to figure out how to fasten it and there was no help, so I figured out my own way of doing it and that's sort of how the whole thing got started." Tom, who owned a furniture store at the time, began to incorporate into his business fasteners that would keep furniture from falling during quakes. The fastener business continued to grow, and in 1994, after the Northridge earthquake, Q-Safety was doing so well that Tom sold his furniture business and began focusing only on fastener sales. His products became a popular item for both household and office safety.

> IN ORDER TO FIND WORK IN THIS INDUSTRY, YOU MUST FIRST FIND THE COMPANIES THAT FOCUS ON EARTHQUAKE-RELATED SERVICES. THE BEST PLACE TO DO THIS IS THE YELLOW PAGES OF ANY MAJOR CITY WHERE THERE IS EARTHQUAKE ACTIVITY, SUCH AS SAN FRANCISCO, LOS ANGELES, SEATTLE, OR PORTLAND, OREGON.

Although there is a definite need for furniture fasteners on the West Coast, Tom says the real reasons for his company's success were beyond his control. "It was the earthquake that grew the business. If we hadn't had any more earthquakes after Whittier, I don't think the business would exist. Or it would exist as a very small accessory to another business. With each earthquake, the business and the industry gets bigger. I don't consider the trauma and destruction that goes along with earthquakes to be anybody's luck, but it's just a fact that there have been earthquakes that have formed this earth for millions of years. They're not going to stop, and people have to take responsibilities for their homes and their work environments to keep them safe."

What do you do all day?

Tom spends most of his time with institutional sales. Because he works with companies mainly by referral, there is very little cold-calling or pavement pounding involved. Usually, a sale will require that Tom put together an estimate for a company that calls to inquire about the products. In order to do this, he must find out how many computers, bookshelves, copiers, and other special equipment the company has. This can usually be done over the phone, but sometimes requires that Tom go out to an office building.

When he is not putting together a sale, Tom is more than likely working with other aspects of the company, such as preparing or sending out catalogs or other information about Q-Safety products. He also spends a lot of time with general bookkeeping, ordering materials, or overseeing some aspect of the manufacturing. At times, Tom works on developing another product. "After the Northridge earthquake I had a lot of people call in and say, "My big-screen television fell over. What do you have for that?" Well, we didn't have anything, so I developed a very long strap because big-screen televisions have a large housing on the back, so they have to be farther away from the wall."

Where do you see this job leading you?

Tom says he's happy right where he is and doesn't expect to make any career changes in the future.

97. Green Architect

description: Much like any architect, a "green" architect oversees the design and construction of a building. However, a green architect also pays close attention to the role a structure will play in its environment. This includes anything from using construction materials that won't contribute to landfills, to making sure that the paints or other products don't contain chemicals that would harm the environment. In addition, green architects consider the way the temperature will be controlled in a building, and try to incorporate energy-efficient avenues for heating and cooling. Doing this involves making additions to both the structure and its surroundings. Green architects may also be involved in creating ways for buildings to promote alternate forms of transportation by allowing for bike storage space. Many green architects are interested in the design and structure of a building, but all green architects are concerned with the future of the environment.

salary: The size of your paycheck will usually depend on the size of the firm for which you work, but in general, a beginning salary won't be below $20,000 a year. Salaries will grow with years of experience, so that someone with eight to 10 years of experience could make as much as $70,000 a year.

qualifications: Anyone going into architecture must have at least a bachelor's degree. However, if your focus is going to be on green architecture, you may want to couple that degree with courses in environmental studies or something similar.

prospects: While the job prospects for architects may depend on the state of the economy, jobs for green architecture are more likely to depend on your entrepreneurial spirit. Environmental architecture isn't always the focus of the firms hiring, but it is always possible to find freelance work that will allow you to address environmental concerns.

characteristics: People working in this field are generally creative and organized at the same time. But perhaps most importantly, they are dedicated to preserving the environment.

Richard Miller *is a green architect.*

How did you get the job?

Richard Miller grew up in upstate New York, where green trees and grass went on for miles. Just as distinctive as his memories of the beautiful landscapes of his childhood is the memory of a hotel chain wanting to build a hotel on the site of the park. "I saw what could happen that would destroy a very beautiful area very quickly," Richard says. It is this particular memory that keeps him dedicated to his decision to preserve the environment in his day-to-day work as a green architect.

Richard works for an architecture firm that focuses on preserving older buildings. However, he also picks up freelance jobs that concentrate on green architecture. Some of his jobs

involve going into a house where one of the family members is chemically sensitive. He will choose paints and insulation that will allow this person to live a healthy lifestyle. Other jobs may require that he design a structure that will be heated by solar power. Richard says that while his educational background (he has a bachelor's degree in architecture) contributed to getting him work at architecture firms, it is the contacts that he has made both within and outside the firm that have allowed him to pick up freelance work as a green architect.

What do you do all day?

In an average day, Richard spends most of his time reviewing drawings and producing details. Because the drafting and drawings are done on the computer, this means that he spends a lot of time sitting at a desk, working on the computer. Although drawing is most of what Richard does, he also spends a great deal of time coor-

THE BEST WAY TO FIND WORK AS AN ARCHITECT IS TO EXHIBIT YOUR SKILLS. THIS CAN BE DONE THROUGH AN INTERNSHIP OR SUMMER JOB. YOUR UNIVERSITY WILL PERHAPS BE YOUR BEST RESOURCE WHEN SEARCHING FOR THESE TYPES OF JOBS.

dinating mechanical, structural, plumbing, and electrical consultants on specific jobs. "The architect is sort of the organizer, and that can really be strenuous," Richard says. Because the

firm he is employed by focuses mainly on preserving older buildings rather than constructing new ones, Richard is always working within the confines of a structure that already exists. Richard faces this challenge with fervor, but it is not the greatest one that his work presents to him. "I think the hardest challenge is to create a community that wastes the least amount of energy and doesn't create any pollution or doesn't destroy the environment at all. It's almost impossible, but I think it is possible, and it is being achieved. That's the excitement of it for me. It is a wonderful challenge," Richard says, adding that this is what keeps him interested in his work.

Where do you see this job leading you?

Richard says that he is currently enjoying the stability of working for an established firm, but that he would eventually like to start up his own company that will focus solely on green architecture.

98. Green Industrial Designer

description: "Green" industrial designers take an ecological approach to the design of resource-efficient products and services. They combine artistic ability and practicality with an environmental consciousness in designing products and services. As corporate America and small businesses come under public scrutiny regarding their environmental impact, they turn to green industrial designers to produce programs, services, and products that not only serve their consciousness but also improve their social image. It falls to the green industrial designer to conjure up and create those ideas. Many work for established industrial design firms, while many others are branching out on their own. In general, the designer meets with clients to discuss the ideas, purpose, and general goal of each project. Then it's literally to the drawing board. Staying on course with costs, impact, and resources, it's up to the designer to come up with the perfect solution for the client.

salary: According to the Industrial Design Society of America, salaries start at around $30,000.

prospects: Overall prospects for a career in industrial design are very promising. The IDSA says that the field is experiencing much growth. According to the IDSA, there are very few people making a living solely as green designers. However, environmental safety concepts are generally incorporated into the overall work. Those who consider themselves green industrial design specialists have a bit of an edge in some markets. Many with this interest are creating their own jobs within large design firms, while others are finding success independently as they create their own companies.

qualifications: Many large corporations and consulting or design firms may require an undergraduate degree in industrial design.

characteristics: The most important qualities a green industrial designer must have are attention to detail, self-discipline, perseverance, an enjoyment of learning, and the ability to work collaboratively.

Wendy Brawer *is a green industrial designer.*

How did you get the job?

Wendy Brawer doesn't have a degree in industrial design. She started out as an artist and sculptor, but says she always dabbled in design work. A trip to Bali piqued her environmental interest, and she returned with a decision to keep an ecological approach to her career. She enrolled in a continuing education class on industrial design, where she learned the nuts and bolts of the profession. She was a quick study; two years later she was teaching a course in green industrial design.

Wendy never worked for a commercial industrial design firm. Instead, she ventured out independently with two hot ideas. Those successful endeavors led to more and more jobs. First she created "The Stamp Out Junk Mail Kit," a packet of information, postcards, and

THERE ARE MANY HELPFUL RESOURCES FOR JOBS. CHECK THE NEWSSTANDS FOR MAGAZINES THAT TARGET ENVIRONMENTAL ENTREPRENEURS AND DESIGNERS. SOME HAVE SPECIAL REGULAR SECTIONS ON ECO-DESIGN.

stickers that enables individuals to control the unwanted advertising mail that is sent to them. She followed that success with the "Green Apple Map," a comprehensive guide to ecologically interesting places in New York City. It's now in its third edition, and has been featured in more than a dozen magazine articles and on television and radio. That led to her work with a large Manhattan housing project, where she designed a comprehensive environmental information program. That program included an EcoCart—

a mobile information center—as well as promotional materials, events, and posters. She has also designed recycling bins for the Times Square area, and has consulted on the environmental impact of products in development.

What do you do all day?

Wendy is her own boss, a one-person operation working out of her home office. Optimally using her time is the key to success, she says. "The need to be careful with your time to stay on top of things is critical." She says that being a member of trade organizations is also an important part of her job, because these contacts are crucial for networking. Much of her day is spent on the phone, networking and talking with colleagues and clients. She also writes proposals and articles, and draws, researches, and designs current projects. Many days are filled with preparing lectures for schools, conference seminars, and workshops, as

being a green industrial designer is a new career many are anxious to hear about. She also, of course, spends time letting the creative juices flow, drafting and sketching ideas, but away from the computer. Wendy says she spends just as much time creating new ideas the old-fashioned way, at the easel, as she does on the computer.

Where do you see this job leading you?

Wendy is headed as far as her ideas can take her. Currently, she's reshaping one of her more popular ideas for the Internet. Her "Green Apple Map" is in such demand that she's working on a way for people to create their own Green Maps for other cities and communities.

99. Green Interior Designer

description: Green interior designers help create environmentally responsible residences and commercial structures. This is a relatively new specialty, as more architectural firms seek to create and renovate space in a manner that reduces impact on the environment. They do this by using products and designs that are energy efficient and ecologically sound—like installing special lighting systems, or using wall insulation made of non-toxic components. They might choose to use natural or recycled materials or one of many other options. However, to be successful, the challenge is to use this green philosophy in a way that meets traditional requirements. Designs must be functional, aesthetically pleasing, within budget, and on schedule. Green interior designers might incorporate environmental aspects into just one element of design, or into the entire construction.

salary: Salaries for this specialty are not much more than for traditional architects or interior designers. Starting salaries are about $22,000–$25,000 per year.

prospects: The prospects for this new style of architecture and interior design are very good. It is a trend that many architectural firms are incorporating. Those with a special interest and experience in environmentally sound design and architecture can expect to do well in this field. There are also good prospects for environmental researchers hired by architectural firms.

qualifications: The obvious route is to head towards interior design and architectural school. Environmental science courses are also recommended.

characteristics: Beyond the general creativity required in doing interior design, this specialty demands an interest in researching new ground. It takes an adventurous personality to fashion new approaches to traditional design.

Kirsten Childs *is a green interior designer in New York City.*

How did you get the job?

It was one particular project that changed how Kirsten Childs and other members of her firm perceived architecture. The Natural Resources Defense Council hired them to renovate a downtown Manhattan warehouse into new office space. The stipulation was that the renovation had to include only environmentally sound and energy-efficient products and design.

That was in 1988, when nobody had heard about "green architecture," and the products were limited. They successfully met the challenge. The firm, Croxton Collaborative, under the direction of Randy Croxton and Kirsten, followed up by incorporating green design into all of its projects. One of the

> ONE TOP GREEN INTERIOR DESIGNER RECOMMENDS THAT, IF YOU ARE ALREADY WORKING IN THE FIELD, YOU SHOULDN'T TRY TO BECOME A SPECIALIST OVERNIGHT. SHE SUGGESTS THAT YOU START MODESTLY WITH YOUR NEXT PROJECT, AND FOCUS ON CREATING ONE ENVIRONMENTALLY RESPONSIBLE FEATURE. FOR EXAMPLE, RESEARCH AND DESIGN AN ADVANCED, ENERGY-EFFICIENT LIGHTING SYSTEM. FOLLOW UP ON THE NEXT PROJECT WITH EVEN MORE FEATURES.

most noteworthy ones was the total renovation and restoration of Audubon House in Manhattan—the headquarters of the National Audubon Society. It is a showcase and a model of an energy-efficient and environmentally responsible workplace.

What do you do all day?

Unlike some other interior designers, Kirsten is involved with a project from the very start. Directing a team of architects and designers, she'll begin to outline plans.

Some features might include advanced energy-efficient lighting systems with occupancy sensors that automatically turn off lights in unoccupied areas. She might direct the use of special wall insulation made of nontoxic compounds, special windows, products made of recycled materials, safe low-toxic interior paints, or natural, undyed carpets that are tacked down to avoid the use of glues.

A typical week, she admits, can be chaotic. Because projects are located all over the world, there's quite a bit of travel involved. Kirsten's also in demand as a speaker at many conferences and seminars because of her innovative and successful design reputation.

Kirsten says architects should keep the pressure on manufacturers to research and create new, ecologically sound construction products. She says that the everyday challenge is to create environmentally conscious design that is both practical and affordable.

Where do you see this job leading you?

Kirsten is now regarded as being at the top of her field. Every project is a challenge, and new ecologically safe construction products are being introduced everyday.

100. Sourcing Manager

(ENVIRONMENTAL COMPLIANCE AND PROTECTION)

description: This is a variation of a job traditionally found in large industry, with job titles like sourcing managers, purchasing managers, or purchasing agents. These new sourcing managers have an added environmental safety element to their jobs; they prevent their companies from making purchasing decisions that would result in harming the environment. Sourcing or purchasing managers are in charge of selecting, ordering, and buying the equipment and materials that the company needs to manufacture its products. While companies must comply with government environmental regulations, there is a way to take environmental protection a step further. An environmental specialist has the knowledge to look at overall projects and purchasing orders and catch potential hazards, ultimately saving companies money in cleanup bills or fines. In many large industries, this job can be a bridge between an environmental compliance department manager and the purchasing officer.

salary: Sourcing managers can expect to earn $35,000 to $55,000 per year.

prospects: It's a traditional job with a new twist. Prospects should be good to create these positions in big industry.

qualifications: One sourcing manager tells us that a business degree isn't necessary, but recommends that interested candidates have a science or environmental background. He says the mechanics of purchasing can be learned on the job. A regulatory or environmental compliance background also helps.

characteristics: One must have good interactive and communication skills to perform well in this career. The job requires an analytical mind to size up situations and projects and to offer recommendations. It also requires patience and diplomacy, because change isn't always readily accepted by some longtime employees.

Aubrey Smith *is a sourcing manager.*

How did you get the job?

With a degree in biology and concentrated studies in microbiology, Aubrey Smith's first job out of school was an internship arranged through the Environmental Careers Organization. ECO is a highly respected national organization that places students in environmental internships, many of which are paying ones. Aubrey's job was a research assistant position with one of the world's largest petroleum companies, where he tried to isolate organisms that would naturally break down oils at time of spills or seepage.

A headhunter helped Aubrey move out of the oil industry and into a job with another company, where he worked in the environmental compliance office.

It was through his own initiative and determination that he proposed and successfully created the job he now has as a sourcing manager (or purchasing agent), with a specialty in environmental concerns.

What do you do all day?

Aubrey's job is to prevent the company from making the wrong decisions in terms of environmental health. That in turn saves the company money and protects the environment.

His task is to evaluate projects from an environmental perspective. He'll pose questions like "Are we jeopardizing or compromising our environmental integrity by trying to save 'x' number of dollars? Will this decision cost us money in the end by having to do environmental cleanups?"

Projects that contain potential environmental implications land on his desk for review every day. He examines all aspects of the project, identifies its implications, and produces a plan that will eliminate or mediate its environmental impact.

"My role is to evaluate chemicals from an environmental standpoint . . . to make sure we're not purchasing things that are upsetting our program. Now we have a better chance to make sure that we don't cause an upset in our compliance program by understanding what changes are going to be made *before* they are made."

Aubrey spends much of the day on the phone or in meetings. In terms of purchasing, he keeps a constant dialogue with the managers and engineers at the six plants he is in charge of. In addition to making sure that supplies are in stock, he also hires contractors.

Where do you see this job leading you?

Aubrey works out of corporate headquarters, but says he would eventually like to be in charge of one entire plant as a plant manager or materials manager. "It would give me an opportunity to play a role in positively impacting the environment—hands on—at a particular plant."

> **REGULATORY EXPERIENCE HELPS YOU MOVE INTO THIS FIELD, BUT YOU DON'T HAVE TO WORK FOR THE EPA TO GET REGULATORY EXPERIENCE. LOCAL GOVERNMENT ENVIRONMENTAL PROTECTION AGENCIES ARE A GOOD PLACE TO GET THAT EXPERIENCE. YOU MIGHT ALSO TRY APPLYING FOR A JOB IN THE ENVIRONMENTAL COMPLIANCE DEPARTMENT OF A LARGE INDUSTRY.**

RESOURCES

Some Major Nonprofit Environmental Organizations

Environmental Defense Fund
257 Park Ave. South
New York, NY 10010
(212) 505-2100

Greenpeace USA
1436 U St. NW
Washington, DC 20009
(202) 462-1177

National Audubon Society
950 Third Ave.
New York, NY 10022
(212) 832-3200

National Wildlife Federation
1400 16th St. NW
Washington, DC 20036
(202) 797-6800
This organization publishes the Conservation Directory, *which lists all wildlife organizations in the United States and is a good resource for those seeking environmental jobs.*

Nature Conservancy
1815 N. Lynn St.
Arlington, VA 22209
(703) 841-5300

Sierra Club
730 Polk St.
San Francisco, CA 94109
(415) 776-2211

U.S. Public Interest Research Group
215 Pennsylvania Ave. SE
Washington, DC 20003
(202) 546-9707

Wilderness Society
900 17th St. NW
Washington, DC 20006-2596
(202) 833-2300

World Wildlife Fund
1250 24th St. NW
Washington, DC 20037
(202) 293-4800

Government Jobs

Bureau of Land Management
Department of the Interior
Washington, DC 20240
(202) 452-5120
The BLM publishes a pamphlet entitled "Career Opportunities in the BLM."

Bureau of Reclamation
Personnel Office
Engineering and Research Center
Building 67
Denver Federal Center
Denver, CO 80225
(303) 236-6914
This agency oversees the construction of major dams.

National Park Service
Division of Personnel Management
Department of the Interior
P.O. Box 37127
Washington, DC 20013
(202) 208-5093

National Weather Service
National Oceanic and Atmospheric Administration (NOAA)
Department of Commerce
Washington, DC 20230
(202) 377-2985
For information, contact your local weather service, which is usually located under U.S. Government, Department of Commerce, National Oceanic and Atmospheric Administration in the phone book. NOAA oversees studies of oceans, the atmosphere, and the space environment. The agency also oversees the National Weather Service. NOAA employs professionals in meteorology, cartography, geography, geology, mathematics, and physics.

Sustainable Agriculture Network
USDA Agriculture Library
National Agriculture Library
1301 Baltimore Blvd.
Room 304
Beltsville, MD 20705
(301) 504-6559
The network can provide you with an Internet address to get job information or answers to specific agriculture questions.

U.S. Army Corps of Engineers
Department of Defense
Personnel and Employment Service
Room 3 D727
The Pentagon
Washington, DC 20310-6800
(202) 761-0660
The Army Corps of Engineers oversees engineering projects on rivers, harbors, and other waterways. One of this agency's top priorities is the protection and restoration of our country's wetlands.

U.S. Fish and Wildlife Service
Department of the Interior
Office of Personnel
Washington, DC 20240
(202) 208-4646

U.S. Geological Survey
Recruitment and Placement
215 National Center
12201 Sunrise Valley Dr.
Reston, VA 22092
(703) 648-6131
The USGS conducts unbiased scientific studies and surveys of natural resources. It is the largest employer of geologists in the United States. In addition to geoscientists, the USGS also employs a number of cartographers, engineers, hydrologists, and others in related fields.

Professional and Trade Associations

Air and Waste Management Association
P.O. Box 2861
Pittsburgh, PA 15230
(412) 232-3444

American Association of Botanical Gardens and Arboreta
786 Church Rd.
Wayne, PA 19087
(610) 688-1120

American Association of Zoological Parks and Aquariums
Oglebay Park, Route 88
Wheeling, WV 26003
(304) 242-2160

American Geological Institute (AGI)
4220 King St.
Alexandria, VA 22302
(703) 379-2480
This is an umbrella organization consisting of 29 geological societies, including the Geological Society of America and other more specialized geological associations. The AGI offers many publications, and the most comprehensive geology-related job listings in the U.S.

American Horticultural Society
7931 E. Boulevard Dr.
Alexandria, VA 22308
(703) 768-5700

American Industrial Hygiene Association
P.O. Box 8390
345 White Pond Dr.
Akron, OH 44320
(330) 849-8888

American Institute of Chemical Engineers
345 E. 47th St.
New York, NY 10017
(212) 705-7338

American Institute of Chemists
7315 Wisconsin Ave. NW
Bethesda, MD 20814
(301) 652-2447

American Meteorological Society
45 Beacon St.
Boston, MA 02108
(617) 227-2425

American Society for Horticultural Science
113 S. West St., Suite 400
Alexandria, VA 22314-2824
(703) 836-4606

American Water Works Association
6666 W. Quincy
Denver, CO 80235
(303) 794-7711
This association provides information about many types of water issues and concerns. Individual professionals are members, as are large metropolitan water utilities and small independent water providers.

AWWA offers many publications and journals, some of which include job opportunities. Annual conferences and chapter meetings are held throughout the United States. Membership is $80 per year.

American Wind Energy Association
122 C St. NW, Fourth Floor
Washington, DC 20001
(202) 383-2500

Association of Energy Engineers
4025 Pleasantdale Rd.
Atlanta, GA 30340
(404) 447-5083

Association of Ground Water Scientists and Engineers
6375 Riverside Dr.
Dublin, OH 43017
(614) 761-1711

Association of Wetlands Managers
P.O. Box 269
Berne, NY 12023

Committee for Sustainable Agriculture
P.O. Box 838
San Martin, CA 95046
(408) 778-7366
This non-profit group offers training primarily for farmers, farm advisors, and those in related fields. Membership fee is $35.

Geological Society of America
3300 Penrose Pl.
P.O. Box 9140
Boulder, CO 80301
(303) 447-2020

Independent Organic Inspectors Association
Rt. 3, Box 162-C
Winona, MN 55987
(507) 454-8310
email: jriddle@luminet.net

Industrial Design Society of America
1142 Walker Rd.
Great Falls, VA 22066
(703) 759-0100
The society offers a career information packet for $10.

The International Marine Animal Trainers Association
1720 South Shores Rd.
San Diego, CA 92109
They prefer written requests for information.

National Association of Environmental Professionals
5165 MacArthur Blvd. NW
Washington, DC 20016
(202) 966-1500
The association offers job listings in a monthly magazine for members only. Annual membership $95.

National Society of Fund Raising Executives
1101 King St., Suite 700
Alexandria, VA 22314
(703) 684-0410
Chapters are located nationwide. The organization offers certification, and training seminars. Career information is offered for a $20 charge. Yearly membership is $170 for regular membership, plus local chapter dues that vary across the country. Membership includes a quarterly newsletter, directory of members, opportunity for certification, yearly conferences, and training seminars.

Environmental Industry Associations
4301 Connecticut Ave. NW, Suite 300
Washington, DC 20008
(202) 244-4700

Society of American Foresters
5400 Grosvenor Ln.
Bethesda, MD 20814
(301) 897-8720
Student membership dues are $25, although dues for other groups may vary. The society offers a career information packet that includes a list of accredited schools and other information regarding careers in this field.

Society of Ecological Restoration
1207 Seminole Highway, Suite B
Madison, WI 53711
(608) 262-9547
This group of professional restoration specialists includes landscape architects, urban planners, and backyard gardeners. The $29 annual membership offers quarterly newsletters and discounted rates for annual conferences. Several journals are also offered at additional costs.

Soil Science Society of America
677 S. Segoe Rd.
Madison, WI 53711
(608) 273-8080

Helpful Publications

ATTRA
Appropriate Technology Transfer for Rural Areas
P.O. Box 3657
Fayetteville, AR 72702
(800) 346-9140
ATTRA publishes an internship opportunities list. This publication can help you find an apprenticeship on a farm.

Careers for Environmental Types & Others Who
    Respect the Earth
Jane Kinney, Michael Fasulo
VGM Careers Horizon
NTC Publishing Group
4255 W. Touhy Ave.
Lincolnwood, IL 60646-1975
(847) 679-5500

Environmental Career Directory
Visible Ink Press
835 Penobscot Building
Detroit, MI 48226
(800) 776-6265
In this directory, found in bookstores, top industry professionals offer their perspectives and advice on environmental careers and the job market. The directory also contains a comprehensive list of trade and professional associations and employment organizations in a job opportunities databank, and offers lists of career resources and publications.

Environmental Opportunities
Sanford Berry, Editor
P.O. Box 788
Walpole, NH 03608
(603) 756-4553
This publication lists jobs in all types of environment-related work, including administration, ecology, fisheries, wildlife, environmental education, natural resources, horticulture, and nature centers, as well as internships and seasonal work. It can be found in the reference section of larger libraries.

Fund-Raising Management Magazine
224 Seventh St.
Garden City, NY 11530
(516) 746-6700
(800) 229-6700
$58 for 12 issues. This publication focuses on issues of interest to professional fund-raisers.

In Business: The Magazine for Environmental
    Entrepreneuring
J.G. Press
419 Stale Ave.
Emmaus, PA 18049
(610) 967-4135

Occupational Outlook Handbook
Bureau of Labor Statistics
441 G St. NW
Washington, DC 20212
(202) 523-1327
The handbook contains articles on and profiles of various occupations, with educational requirements and job prospects.

Miscellaneous

ACCESS/Networking in the Public Interest
96 Mount Auburn St.
Cambridge, MA 02138
(617) 354-9458
ACCESS offers a publication listing job opportunities in many fields.

Global Environmental Fund
L.P. GEF Management Corporation
1250 24th St. NW, Suite 300
Washington, DC 20037
(202) 789-4500
The fund invests in companies whose products or services help improve the environment.

Internships
Environmental Careers Organization
286 Congress St.
Boston, MA 02210
(617) 426-4375
ECO is probably the most respected environmental internship program in the United States. The organization places undergraduate and graduate students in internships that last anywhere from three months to two years. They have regional offices throughout the country. They also publish The Complete Guide to Environmental Careers.

Island Press
1718 Connecticut Ave. NW
Washington, DC 20009
(800) 828-1302
(202) 232-7933
This is a nonprofit publisher of environmental books.

ABOUT THE AUTHOR

*Debra Quintana is a television news reporter living in New York City
with her husband and four-year-old daughter, Alexandra.*